Library of
Davidson College

The Science of Musical Sounds

This is a volume in the Academic Press Series
COGNITION AND PERCEPTION

Series Editors
E.C. Carterette
M.P. Friedman

The Science of Musical Sounds

Johan Sundberg
The Royal Institute of Technology
Speech Communication and Music Acoustics
Stockholm, Sweden

ACADEMIC PRESS, INC.
Harcourt Brace Jovanovich, Publishers
San Diego   New York   Boston   London   Sydney   Tokyo   Toronto

This book is printed on acid-free paper. ∞

English translation copyright © 1991 by ACADEMIC PRESS, INC.
© 1973, 1978 och 1989, Johan Sundberg och Proprius förlag, Stockholm
All Rights Reserved.
No part of this publication may be reproduced or transmitted in any form or by any means, electronic or mechanical, including photocopy, recording, or any information storage and retrieval system, without permission in writing from the publisher.

Translated from the Swedish title: Musikens Ljudlära: Hurtonen Alstras Och Uppfattas. Translated by Johan Sundberg.

Academic Press, Inc.
San Diego, California 92101

United Kingdom Edition published by
Academic Press Limited
24–28 Oval Road, London NW1 7DX

Library of Congress Cataloging-in-Publication Data

Sundberg, Johan, date
 The science of musical sounds / Johan Sundberg.
  p. cm. -- (Academic Press series in cognition and perception)
 Includes index.
 ISBN 0-12-676948-6
 1. Sound. 2. Music--Acoustics and physics. I. Title.
II. Series.
ML3807.S95 1991
781.2'3--dc20   91-4267
                                                               CIP

PRINTED IN THE UNITED STATES OF AMERICA
91 92 93 94   9 8 7 6 5 4 3 2 1

Contents

Preface **ix**

1. **Introduction**
    I. Music Acoustics 1
    II. The Music Sciences 2
    III. How? and Why? 4

2. **What Is Sound?**
    I. Introduction 8
    II. Simple Oscillations 9
    III. Sound Propagation 12
    IV. Spherical and Plane Waves 16
    V. Doppler Effect 16
    VI. Sound and Obstacles 17
    VII. Sound Intensity Measures 18
    VIII. Decibel and Sound Level 20
    IX. Complex Tones 24
    X. Several Simultaneous Spectra 29
    XI. Beats 30
    XII. Resonance 31
    XIII. Attenuation 33
    XIV. Standing Waves 36

3. **Ear and Hearing**
    I. Introduction 38
    II. Anatomy of the Ear 39
    III. Hearing 44

4. Scales, Tunings, and Temperaments

    I. Introduction   78
   II. Some Words on the Art of Computing   78
  III. What Is an Interval?   80
  IV. Calculating Intervals   81
   V. Consonance and Dissonance   83
  VI. Pythagorean Tuning   86
 VII. Just Tuning   87
VIII. Equally Tempered Tuning   89
  IX. Cent   92
   X. Comparing Tunings   94
  XI. Calculating Intervals   97
 XII. Demands on a Scale   98
XIII. The Scale in Practice   100
XIV. Craving for Stretching   103

5. Wind Instruments

   I. Introduction   106
  II. Tube Resonators   108
 III. No-Feedback Instruments   116
 IV. Feedback Instruments   128

6. String Instruments

   I. Introduction   141
  II. The String   141
 III. Instruments with Struck Strings   145
 IV. Instruments with Plucked Strings   152
  V. Instruments with Bowed Strings   154

7. Rod and Membrane Instruments

   I. Rod Instruments   161
  II. Membrane Instruments   164
 III. Bells and Gongs   167

8. Room Acoustics

   I. Introduction   169
  II. Reflection   169
 III. Absorption   172
 IV. Calculating Reverberation Time   174
  V. Modulation Transfer   177

Contents

      VI. Timetable for Reflection   179
     VII. One Sound per Ear Please!   181
    VIII. Podium Acoustics   182
     IX. Self/Others Balance   183
      X. Music Composition and Room Acoustics   184
     XI. A Personal Reservation   186
     XII. A Final Remark   186

9. Music and Electronics

       I. Introduction   188
      II. Electroacoustic Conversion   189
     III. A Quick Look at the World of Synthesizers   200
     IV. Acoustic Measurement Equipment   205
      V. Some Words about Research Methods   210

10. Music as Communication

       I. Introduction   212
      II. Tone Glue   214
     III. A Sound Means So Much!   219
     IV. Performance Rules   221
      V. Bibliography   226

References   229

Index   231

Preface

A long time has passed since the first time I wrote this text; the first edition was printed in 1973 and the third 1989. During these years there has been a substantial development of knowledge within the field of music acoustics. In particular, the domestication of computers has entailed great gains. For this reason the author felt it as a great privilege to revise the book. With regard to the development of knowledge and understanding in the area of music acoustics we are living in a wonderful era!

Earlier, music acoustics had to deal with rather basic problems, such as musical intervals between pure sinetones, characteristics of stationary spectra of instrument tones and other aspects of music that were far remote from the essence of experiencing music. Now musically much more burning topics can be attacked. This means that music acoustics has now the potential of fascinating a great number of musicians and other musically interested persons.

As a sign of the growth of the topic, a number of basic textbooks have appeared in recent times, starting in 1969 with J. Backus, *The Acoustical Foundations of Music;* A. H. Benade, *Fundamentals of Musical Acoustics;* D. Hall, *Musical Acoustics: An Introduction;* J. H. Pierce, *The Science of Musical Sound;* T. D. Rossing, *The Science of Sound;* and most recently, N. H. Fletcher and T. D. Rossing, *The Physics of Musical Instruments.* There is no need for another contribution from this author to this impressive collection of standard textbooks.

Rather, the author felt that a simple, popular book presenting a survey might be needed, a book that could serve as a first appetizer for musically interested people who do not want to learn more anecdotes about composers but who rather are curious to learn about new aspects of music. This book is written for such people. Where physics or mathematics appear in the text, explanations are also presented. Thus, the author's ambition and hope has been that basically everyone who is musically interested and curious to know more can enjoy reading the book.

In writing the text, I have received help from my friends and co-workers in the Music Acoustics Group at the Department of Speech Communication and Music Acoustics of the Royal Institute of Technology in Stockholm, who have read some of the chapters and offered remarks. I am also indebted to Professor Edward Carterette, editor of the series in which the book appears, for good advice and for convincing me to translate the book.

Johan Sundberg

CHAPTER 1

Introduction

I. MUSIC ACOUSTICS

What kind of an animal is music acoustics in the disparate world of science? What does it want to achieve? What does it have to offer? These are questions that this book will answer. But at this point it might be worthwhile to sketch at least the gross contours of the answers.

Music acoustics is both old and young as a science. Early contributions were made back in the Greek Antiquity when Pythagoras wondered why certain musical intervals emerged from certain string length ratios. In our time an early important key figure was a German physician and physicist in Potsdam, Hermann von Helmholtz. In 1862 he published a thick book with a thought-provoking title, *Die Lehre von den Tonempfindungen als physiologische Grundlage der Musiktheorie (The Science of Tone Perception as a Physiological Basis for Music Theory)*. The gospel implied by this title is that music theory ultimately goes back to how our sense of hearing perceives the tones (i.e., that music theory can be based on the theory of auditory perception). This view has been preserved by several scientists in the music acoustics field, and it is frequently supported by research results, also in our time. Therefore the characteristics of our sense of hearing is an important part of music acoustics.

Hermann von Helmholtz dealt not only with the sense of hearing, but also developed theories for several music instruments that explained how they worked (i.e., how the tone is generated in the instrument). Many of his theories have not been greatly revised by later research. His descriptions of, for example, the flue organ pipe or the vibration of the bowed string are still valid. The development of theories describing the function of the various music instruments is regarded as one of the major tasks in music acoustics research. Indeed, it is a necessary task in our time. In many countries it is carried out in collaboration with the music instrument builders; they obviously need to understand how the things they manufacture function. (For comparison, just imagine the situation in which car brake manufacturers did not really understand why the brakes reduced the speed of the cars and just tried to make the brakes in the same way as their fathers did!)

In the 1930s the use of electronic technology was started in music instrument research. An American researcher of Swedish extraction, Carl Seashore, got hold of a so-called tonometer. This device allowed him to examine in detail music sounds from an acoustical point of view. He headed a research group in Iowa, which enthusiastically and productively studied performed music (e.g., intonation, tone duration, and other tone properties that musicians determine when playing). Seashore published the results in another classical book in music acoustics, *The Psychology of Music.*

Another page in the history of music acoustics was turned in the 1970s when the digital computer was domesticated, becoming an obedient research tool. This meant that much more difficult questions could be tackled than before. For example, it was possible to calculate how different bore shapes in wind instruments affected tuning and tone color. As this and other, more difficult problems could be solved, the possibilities for music acoustics to tell interesting things about music increased.

This nutshell historical survey may perhaps give an idea as to what kind of problems music acoustics used to work with and is working with at present. However, history should not determine what scientific research should be dealing with, although it often so happens, unfortunately. Rather, the questions that develop from the research results should determine activities and boundaries.

II. THE MUSIC SCIENCES

We have just seen that music acoustics is not the same as musicology, a discipline that often tends to be more or less equal to music history. In fact, music has given rise to an entire family of music sciences. To find out how music acoustics fits into this family let us start from the beginning by asking "What is music?" This question is particularly significant for the music sciences, because an answer would provide a definition of the research object.

But, unfortunately, the question is difficult to answer. Music may mean so many widely different things: the score, the sound sequences, the experiences in a listener, etc. So what is the research object of the music sciences?

Obviously, it is the music itself. But there is no convincing way of defining music. As a consequence, it is most time-consuming and rarely rewarding to try to reach a common view as to what music really is. In such situations it is often wisest to circumvent such questions with respectful silence: In most cases people know what music is, so let us refrain from trying to find a tenable scientific definition!

II. The Music Sciences

However, there are many tangible manifestations of the fact that music exists, as shown in Figure 1.1. Everything has a history, and music is no exception to this principle. History of music is, because of a strong tradition, the main angle of attack of traditional musicology. There is always an interaction between society and what happens in society, and again, music is no exception. Music sociology works with this aspect of music. Playing and listening to music generates experiences, reactions, behaviors, and cognitive processes, and these are the domain of music psychology. The composer creates a music score, which provides music theorists with their research object. The players generate acoustic signals, which are captured by the listener's ears. These signals are the research object of music acoustics.

Thus we find that another music science has developed around each of the many forms in which music manifests itself: musicology (i.e., music history), music theory, music psychology, music sociology, music acoustics. There are more of them, and the number will certainly increase in the future.

At first sight, some of these angles of attack may seem farfetched. But humans are by nature curious, and apparently, music experience can be enriched and enhanced by knowledge from many different fields. Perhaps the acquaintance with the stories about Beethoven's attitude toward Napoleon adds to the excitement of listening to his famous Third Symphony.

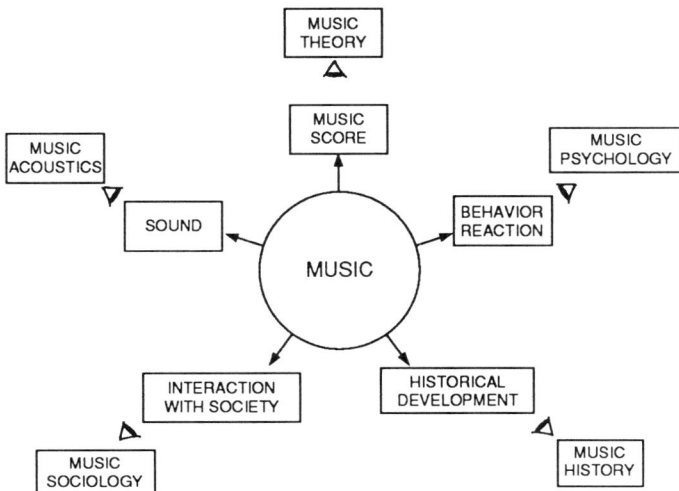

Figure 1.1 The music sciences. Music is manifested and can be observed by the researcher's examining eye from several different angles. Each of these angles has given rise to a music science. Music acoustics is one of them.

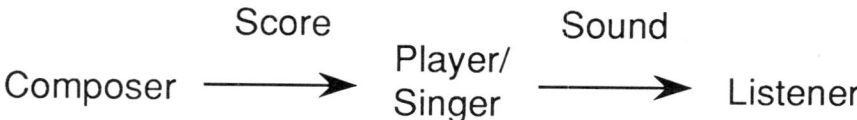

Figure 1.2 Main links in the music communication chain. It often contains more links, such as systems for recording and reproducing sound, and sometimes fewer, such as when the score or the musician is missing.

Music is transferred from composer to listener in the manner illustrated in Figure 1.2.

Note that two of the links in this chain have a particularly close relationship to music *per se:* the score and the sound. Further, although the relevant function of the three persons in the chain are mental processes, the score and the sound are the only links in the chain that are physical objects.

This is significant. It implies that the score and the sound are the only sources of information that can be submitted to a strict scientific analysis according to the model used in natural sciences (i.e., the results need not necessarily be dependent on subjective interpretations). Music acoustics analyzes the sounds and the sound sequences of music by means of the methods used in natural sciences. If essential information about music can be gained in this way, the sounds of music obviously constitute a relevant research topic.

Music acoustics is often called *musical acoustics*. The choice of labels should not be of any appreciable concern to adult brains. However, there is a reason why the name *music acoustics* is used in this book. "Musical" is an adjective referring to a gift that has been given to certain persons. This gift cannot be given to a science. This is one reason to avoid using this term. Further, the other music sciences are called music psychology, music history, music sociology, etc. In analogy with this we should start talking about musical history, musical sociology, musical psychology, etc., which sounds very awkward indeed! Hence the use of the term *music acoustics*.

III. HOW? AND WHY?

What kind of information may be gained by a study of music according to a natural sciences method? The answer must depend on what factors determine the acoustic shape of the music. At first glance we may think that the composer is the dictator who decides everything. Some more reflection will reveal a number of other contributors or contributing factors: the composer's music teachers, what kind of music taste was predominant when the piece was composed, how the

III. How? and Why?

composer felt when composing the piece, and so on. However, there are also a number of factors outside the composer's reach (i.e., the player's understanding of the piece, the acoustic properties of the instrument(s), the work and skill of the instrument builder). All these factors—and there are certainly more than those mentioned here—contribute to the shaping of the sound sequences created when a piece of music is being performed.

Thus, there are several, very different factors contributing to the sound sequences of music. Yet, they all share a common principle. What principle that is may be suggested by the following thought experiment.

Imagine one of the great organ pieces by J. S. Bach played first on a romantic organ from the late 19th century and then played on an authentic instrument. Imagine a piano piece by Debussy, first played on a clavichord and then on a grand piano. Imagine Rimsky-Korsakov's *Scheherazade* played first by a renaissance ensemble with shawms, viols, lutes, etc., and then by a modern symphony orchestra. When listening to the latter alternative in all these cases you are likely to feel an inner "Of course! This is the way in which this piece should be played!" But from where did this certainty originate?

The origin is, of course, not randomly developing traditions or other random factors, which, by the way, represent overly boring research objects. Instead, the origin is inherent in the human system, because music is tailored for the human capacity of perceiving and communicating by sounds. That is the common principle mentioned above, contributing to the shaping of the sound sequences of music. This adaptation to the characteristics of the human system is an overly inspiring research object, by the way.

Already the fact that the composer is a human being implies that the products carry signs of human mental effort. When composing and orchestrating, the composer consciously and unconsciously takes into account how the sounds will be perceived by the human sense of hearing. He or she would ask her- or himself things such as "What will be best here?" and "How should this be orchestrated?" The same applies to a higher or lesser degree also to musicians, instrument builders, and other people contributing to the shaping of the sound sequences. Therefore, the shape of these sound sequences contains explicit and implicit information about humans.

The title of this chapter was taken from a book with the title *Why? and How? A Key to the Natural Sciences* by Brewer, Moigno, and Parville published in Swedish first in 1890. Research following the tradition within the natural sciences should ideally answer these two questions.

The answer to the first question, "How?", is an exhaustive description of the research object. In the case of music acoustics this description does not work with the terminology typically used in music descriptions (e.g., "happily," "mystical," "determined," "light," "dark," "like birchtrees in springtime," and other expressions that attempt to describe what the piece might remind a listener of in a more or less poetic fashion. Rather, descriptions in music acoustics use acoustic terminology. Thus, they provide an acoustic description of the sound waves, containing specifications of frequencies in Hertz, of intervals between scale tones expressed as frequency ratios, of the amplitudes of sine wave components, of tone onsets, and of decays, and other temporal aspects of the sounds. To the typical reader of music books this probably appears as a most bizarre way of talking about music. The advantage is, however, that the terms can easily be defined, so that everyone has a chance to understand what they mean.

Regarding the question "Why?" the answer should explain the description emerging from the "How?" question. Let us think of an example. Why does a piano tone possess precisely its own and no other acoustic characteristics? There are several different factors that contribute. One explanation is the function of the instrument, or in other words, the theory that describes how the tone is created in the instrument. This theory explains how the material and tension in the strings, the properties of the hammer and sound board, etc., contribute to determine the properties of the tone. One important task of music acoustics is to formulate theories describing the sound generation within the instrument.

The question "Why?" also leads to other aspects of music sounds. Explanations for the sounds of music can be found not only in the instrument, but also in the properties of human hearing. The peculiarities of human auditory perception are taken into account both when the instrument is constructed and when it is played. For this reason, theory of auditory perception is another essential part of music acoustics. Of particular relevance is, of course, those types of auditory perception that occur mainly in music (e.g., the pitch, which is categorized into a limited number of scale tones; tone durations, which are categorized into certain stereotypes called *note values;* and the perception of vibrato tones).

When the scope is widened from aspects of a single instrument tone to the sound sequences of a piece of music, things obviously get more complex. Music sounds completely impossible from a musical point of view, when performed exactly as nominally specified by the music score. Musicians make music meaningful by lengthening and shortening tones, by accentuating certain tones, by making some tones louder and others softer, by inserting micro pauses between various tones, etc. By these subtle means the player makes the music interesting to listen to. These aspects of music sounds represent another important part of music acoustics.

When we listen to music it has passed at least one transmission link before it reaches our ears, namely, the acoustics of the room in which the music is played. Further, we mostly listen to recorded music (i.e., it has passed two other links: the sound recording and reproduction systems). Acoustics of music rooms and sound recording and reproduction systems are other important parts of music acoustics.

In this book the reader is invited to share the author's fascination for all these different aspects of music. Hopefully, the reading will provide a generous reward for the work.

III. How? and Why?

Work? This word was not chosen by accident. Music acoustics is not as easy to digest for the typical reader of music books as the usual texts, which speak about music in a more entertaining and anecdotal way. For readers used to texts in the natural sciences area, the difficulties would be nil. The difference between these groups of readers would go back to reading habits. A pertinent piece of advice seems to be similar to what is recommended for safe driving: Never go faster than the reader, the text, and the content permit. Readers who are used to humanistic texts almost always read natural science texts far too fast, and the result then is that the text appears completely impossible to understand. If, however, the reading speed is reduced in accordance with the recommendation above, the same text may become understandable, perhaps easily understandable, and most certainly completely fascinating! With these humble words the author wishes his readers an enjoyable exploration of the science of music sounds.

CHAPTER 2

What Is Sound?

I. INTRODUCTION

The reason why we perceive sound, normally, is that something causes our eardrums to vibrate. There is air on both sides of the eardrums. Therefore it is natural that the vibrations are caused by variations of the air pressure. These variations suck the eardrums outward and press them inward, although by an incredibly small amount. Still, the physical equivalent to sound is air pressure variations.

This important fact, that the physical equivalent of sound is pressure variations, can be visualized in various ways. One way is to put a sheet of paper over a loudspeaker with the cone facing the ceiling and the back side facing the floor. The loudspeaker should be emitting a low-frequency tone, (e.g., hum). Under these conditions one can see that the paper sheet vibrates, jumping up and down. It is the pressure increases that push the paper up, away from the speaker membrane, whereas the pressure decreases suck it downward. In this manner the paper sheet simply follows the pressure variations.

Given the fact that sound corresponds to minute variations of the ambient air pressure, it is natural that the properties of these variations determine the properties of the sound that we perceive: After all there must be some order in the physical world! If the loudspeaker emits a tone, the paper jumps up and down rhythmically and regularly, because a tone corresponds to regular pressure variations. More specifically, the regularity implies that one single pattern of variation repeats itself all the time. This pattern is called a *period*, and the repetitious variation pattern is called *periodical*. The opposite case is *aperiodical* variations, and when we listen to aperiodical variations, we perceive sounds lacking a pitch (e.g., noise).

In music, tones obviously play an important role. Therefore, the periodic variation or oscillation is a cornerstone in music acoustics.

II. SIMPLE OSCILLATIONS

Let us now return to the loudspeaker and ask what, exactly, happens. The membrane of the loudspeaker facing the ceiling moves up and down. When it is moving upward, it chases the air closest to it upward. The air particles thus displaced then displace their upper neighbors, which, in turn, displace their neighbors, and so on.

When the membrane is moving downward, the corresponding procedure is repeated, although with opposite direction: The membrane sucks the air particles closest to it downward, and these particles then suck their upper neighbors down, and so on. In this way the loudspeaker membrane generates motions among the air particles. It is such motions that reach the paper we talked about before and brings it into vibration.

The motion of the loudspeaker membrane generates motion in adjacent layers of air particles. What we need to perceive sound is air pressure variations, as mentioned. How can these air particle movements generate pressure variations? The answer is not particularly complicated. When air particles approach each other, compression results and the pressure rises, and when they part, rarefaction results and the pressure drops. These situations must occur when the loudspeaker membrane is moving. There must be compression of air when the loudspeaker is moving upward and rarefaction in the opposite case.

We are well-acquainted with another type of air pressure variation: Sometimes there is a high pressure and the weather is sunny and fine, and sometimes there is a low pressure and there is rain and wind. This slowly varying pressure is called the *atmospheric pressure.* Such pressure variations can be recorded by means of a barograph. However, they are many, many times greater and slower than those that we perceive as sound. This difference is marked by calling the pressure variations that may generate sound sensations *sound pressure* instead of air pressure. If barographs had been much more sensitive than they are and if the paper they write on was moving much quicker than it is, it would give us a registration of sound pressure variations. If the loudspeaker emits a pure, simple tone, we would obtain a curve such as the one shown in Figure 2.1.

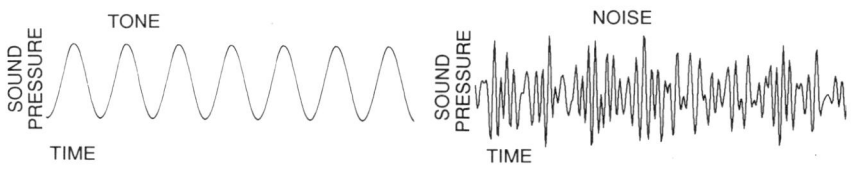

Figure 2.1 Example of periodic (*left*) and aperiodic (*right*) sound pressure variations.

A curve of this kind thus shows how the sound pressure varies with time. We can see that it oscillates regularly and evenly. It varies around a constant average equal to the ambient, atmospheric air pressure just mentioned, which meteorologists measure.

The curve thus shows the sound pressure oscillations around the atmospheric pressure. Such a curve answers three basic questions, and the answers provide an exhaustive description of the periodic oscillation. Let us invent a concrete example.

QUESTION 1: How often is the variation pattern repeated per second?
ANSWER: 100 times per second.

This means that the *frequency* of this tone is 100 hertz (Hz). [Earlier, other terms were used for this unit: cycles per second (cps) or periods per second (pps).] For high frequencies the unit kilohertz (kHz) is frequently used: 1 kHz = 1,000 Hz. The number of periods per second, of course, depends on the length of the period i.e., the *period time*. In other words, frequency is inversely proportional to the period time.

Frequency is decisive to the pitch perceived: the higher the frequency, the higher the pitch. An average listener can perceive sound in the frequency range from 20 to 20,000 Hz, approximately. The keys on a grand piano generally represent tones from about 27 to 3,500 Hz.

QUESTION 2: How much does the sound pressure deviate maximally from the atmospheric pressure?
ANSWER: 0.5 pascal (Pa).

This implies that the maximum deviation, or the (peak) *amplitude,* amounts to 0.5 Pa. One pascal equals a pressure corresponding to the force of 1 newton (N) over 1 m². The amplitude is decisive to the loudness of the tone perceived: the higher the amplitude, the louder the tone. The lowest and highest sound pressure amplitudes that we can hear as tones are between 0.00002 Pa and 20 Pa. A common cousin of the amplitude is the root mean square (RMS) value of the amplitude. It equals the peak amplitude multiplied by $1\sqrt{2} = 0.707$.

QUESTION 3: In which direction is the sound pressure changing at this moment?
ANSWER: The phase is 0 at this moment.

This meaning of the word *phase* is the same as the one used for specifying moon shapes, being another case in which there is a need to specify in what direction something is changing. In many respects, our hearing is not very sensitive to phase information. However, it is quite significant to our ability to tell directions. The phase is given in measures of angle: degrees or radians. One half turn corresponds to $180° = 1\pi = 3.14$. The background to this is illustrated in Figure 2.2.

II. Simple Oscillations

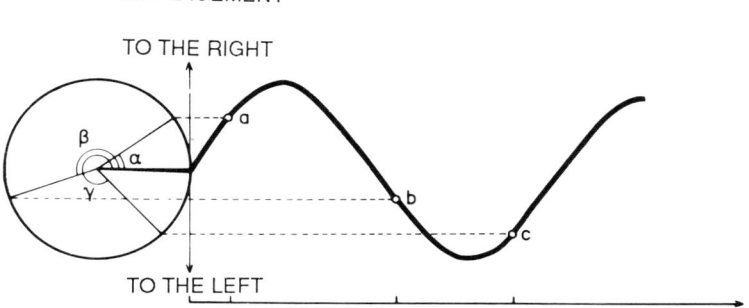

Figure 2.2 Illustration of how a sine wave curve can be described by means of an ever-increasing angle.

In the right part of the figure, we recognize the sound pressure variations from Figure 1.2, even though the time scale here has been expanded. The sound pressure values a, b, and c occur at time t_a, t_b, and t_c. The three values can be given as three positions of a radius that is rotating at a constant speed in the circle shown to the left. These positions can be defined by means of the phase angles α, β, and γ. This is the background to the fact that phase can be specified as an angle.

As our sense of hearing cannot perceive phase in many situations, there is rarely a need for specifying it when one single tone is sounding. It is more relevant when two or more tones sound simultaneously. If two tones of equal amplitude and frequency sound simultaneously, their phase relationship is particularly relevant. If the phase relation is zero, the sound pressure oscillations are in phase, so that at any moment both strive to change the sound pressure in the same direction. The result then is a doubling of the amplitude. If, however, they are in counterphase, i.e., if their phase relation is 180° or half a turn apart, they strive to change the sound pressure by the same amount but in opposite directions at any moment. The result then is zero; they cancel each other. We will return to this phenomenon later in this chapter. In a coming chapter we will also see that the phase relation between the signals reaching the left and the right ears has great relevance to our capability of telling from which direction a sound is coming, as was mentioned.

If one spends a few moments to contemplate the figure, one can also realize that the vibratory motion shown can be mathematically described by means of an equation containing the sine of the ever-increasing phase angle. The course of the sound pressure shown in the figure can be computed from the following equation: the amplitude at time t

$$A_t = K \cdot \sin 2\pi ft \tag{2.1}$$

where f is the frequency and K is a constant that takes care of the task of producing a curve that is of appropriate amplitude for the paper used for the graph. Using the values given in Table 2.1 below, the readers can allow themselves the memorable and unique excitement of constructing their own private sine wave curve. It gives a tangible idea of the reason why acousticians tend to speak about sine tones.

In fact, single sine wave tones are rare animals in the world of acoustics. Soft whistling is our main possibility of generating a fair approximation of a sine tone. Normally sine waves behave like many animals: They go in herds. We will soon return to this issue.

III. SOUND PROPAGATION

In our example with the loudspeaker and the sheet of paper, we saw that the air particles were brought into motion by the movements of the loudspeaker membrane. We also saw that this must cause compressions and rarefactions of the air, which obviously correspond to air pressure variations. When the membrane is moving upward, the layer of air particles resting on it is compressed (i.e., the

Table 2.1 Sine for different angles given in degrees.

Angle	Sine	Angle	Sine
0	0.00		
10	0.17	190	−0.17
20	0.34	200	−0.34
30	0.50	210	−0.50
40	0.64	220	−0.64
50	0.77	230	−0.77
60	0.87	240	−0.87
70	0.94	250	−0.94
80	0.98	260	−0.98
90	1.00	270	−1.00
100	0.98	280	−0.98
110	0.94	290	−0.94
120	0.87	300	−0.87
130	0.77	310	−0.77
140	0.64	320	−0.64
150	0.50	330	−0.50
160	0.34	340	−0.34
170	0.17	350	−0.17
180	0.00	360	0.00

III. Sound Propagation

pressure rises), and this pressure increase propagates to adjacent layers of air particles. Conversely, when the membrane moves downward, a pressure drop results, which propagates to adjacent layers of air particles.

What happens under these conditions is that the air particles are pushed slightly back and forth in the direction of sound propagation. Oscillations of this kind, which happen back and forth in the direction of sound propagation, are called *longitudinal*. It is easy to realize that sound is propagated by means of longitudinal oscillations. For example, the air particles closest to the eardrums must move in this way to move the membranes. Oscillations normal to the direction of propagation also occur. The waves on the water surface are one example. Such waves are called *transversal*.

How is it possible that longitudinal oscillations generate pressure waves? This is illustrated in Figure 2.3. The upper part of the figure shows a row of bullets linked together by small coil springs. As each bullet weighs something, it possesses a certain mass, and the row can obviously be compressed because of the springs. Consequently, this system has properties similar to those of a row of air particles. An air particle is light, but still it must weigh something. A row of air particles can be compressed; what would otherwise be the point of using air in bike tires?

Let us imagine that we start a motion of the left-most bullet. Then phenomena occur that are illustrated in the snapshot shown in the middle part of the figure. At certain locations the bullets are crowded; at other locations they are scarce. The resulting pattern of pressure variation along the line is shown in the bottom part of the figure. This pattern of crowded and scarce bullets does not remain but travels along the line of bullets.

The same situation occurs in sound propagation. A distribution of overpressures and underpressures arises that travels away from the sound source in the direction of sound propagation. This distribution of pressures is called a *sound wave*. It has certain similarities with a water wave, although sound propagates with longitudinal waves whereas the waves on the water surface are transversal waves. But if we take a snapshot of the distribution of overpressures and underpressures along the pathway of propagating sound, the result is similar to the wave pattern on the water surface, as can be seen in the bottom part of Figure 2.3. The length of the sound wave, i.e., the shortest distance between two points with identical phase, is called a *wavelength*. It is usually honored by a special symbol, (λ), which is lower case l in the Greek alphabet.

The sound wave shown in the bottom part in Figure 2.3 is similar to those in Figures 2.1 and 2.2. However, it is extremely important to observe a major difference between them. In the two last-mentioned figures, *time* is the horizontal axis, whereas in Figure 2.3 it is *distance*.

It is somewhat fascinating that the individual air particles move back and forth in the direction of sound propagation, thereby pushing and sucking each other, such that a sound wave is generated in space. If one arranges a row of microphones along the pathway of the sound propagation and records the pressures that these microphones sense at the same instant, one would obtain values according to the sound wave illustrated in the bottom of Figure 2.3. If one would record what one of these microphones senses as a function of time, one would obtain a curve of the type shown in Figures 2.1 and 2.2.

The length of sound waves as well as the waves on the water surface is dependent on the period time: the shorter the period time, the shorter the wavelength. The wavelength also depends on the speed at which the wave propagates.

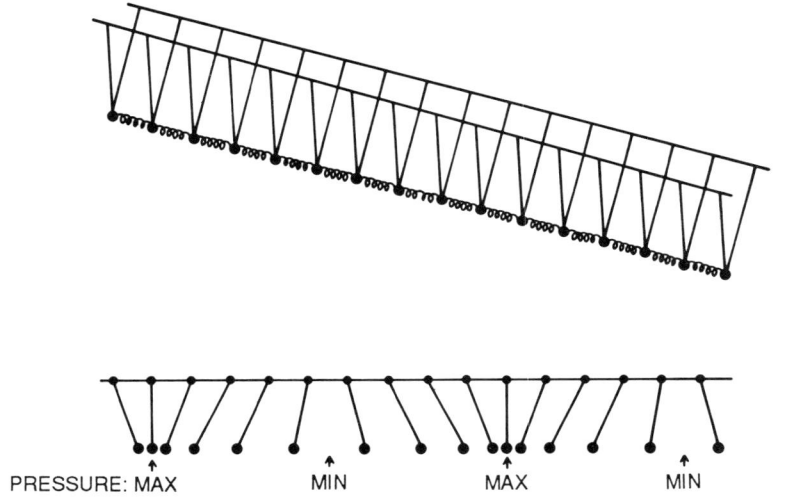

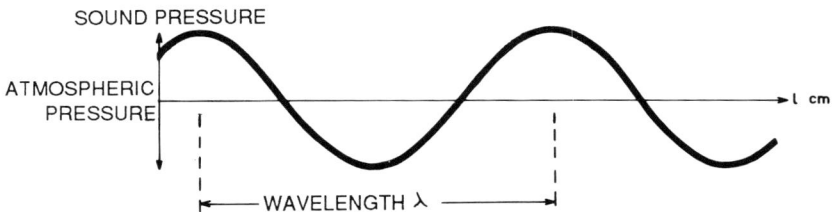

Figure 2.3 Illustration of the principle explaining why movements in air particles generate overpressures and underpressures.

III. Sound Propagation

This speed is called the *speed of sound**, which also is generally honored by a symbol of its own (c). At normal indoor temperatures the speed of sound in air is about 340 m/sec. The relation between the period time T, the speed of sound c, and the wavelength λ is

$$\lambda = c \cdot T \tag{2.2a}$$

This equation can be turned around according to all basic rules for calculation:

$$T = \lambda/c \tag{2.2b}$$

$$c = \lambda/T \tag{2.2c}$$

The last version is perhaps easiest to remember. As a matter of fact, there is law and order in mathematical formulas: The units are always the same on both sides of the sign of equality. Speed is measured in distance per time unit, such as m/sec. Wavelength is a distance, and period time is obviously time. Thus, as long as one can recall that c is alone on one side and that speed is measured in distance per time, then one can never forget that the other side has wavelength over period time.

Some more versions can be generated if we introduce the relation between the period time T and the frequency f, $T = 1/f$:

$$c = \lambda f \quad \text{or} \tag{2.3a}$$

$$\lambda = \frac{c}{f} \quad \text{or} \tag{2.3b}$$

$$f = \frac{c}{\lambda} \tag{2.3c}$$

Equation (2.3a) tells that the speed of sound equals the product of frequency and wavelength: the lower the frequency, the longer the wave. We recognize this relation from many music instruments. The bass strings are long in a piano; the organ pipes for low notes are tall. Low tones need long tools.

The speed of sound in air is about 340 m/sec, as mentioned, which equals about 1,230 km/hr or 770 miles/hr. The speed is affected by the temperature, increasing by about 0.6 m/sec per degree centigrade. As we will see later, this is significant in many connections, e.g., for the tuning of wind instruments.

The speed of sound is mainly dependent on the properties of the medium in which the sound propagates. Table 2.2 shows that these effects are quite great.

---

*The speed of sound propagation is not the same as the so-called particle velocity, which is the speed at which the individual air particles travel when they oscillate. Although speed of sound propagation is a constant for the particular medium or material, the particle velocity varies depending on both amplitude and frequency.

Table 2.2 The speed of sound in various media at 0°C

Medium	Speed (m/sec)
Rubber	70
Oxygen	317
Air	331
Nitrogen	337
Hydrogen	1270
Seawater	1440
Steel	5050

These differing speeds of sound propagation imply that the wavelength for a given frequency varies considerably, depending on the medium in which the wave travels. For instance, the wavelength for a wave propagating in seawater is about 20 times longer than that of a wave traveling in rubber and about 4.4 times longer than that of a wave traveling in air. As we will see in Chapter 6 on string instruments, the speed of sound propagation does not determine the lengths of strings.

IV. SPHERICAL AND PLANE WAVES

Sound is radiated from a sound source such as a loudspeaker not only straight ahead, but in all directions. The ratio between the wavelength and in the case of the dimensions of the sound source, e.g., the radius of a circular loudspeaker is decisive. If the dimensions are small as compared with the wavelength, the source can be regarded as a pulsating sphere, radiating in all directions with the same efficiency, approximately as a light bulb emits light in all directions. This is called *spherical propagation* and *spherical waves,* because the wave front is curved. At a sufficiently great distance from the source, the curving of the wave front is negligible. In such cases the wave is called *plane*. Plane waves also occur in narrow ducts, such as in wind instruments. In the bell of certain wind instruments, the waves are spherical rather than plane. If the dimensions of the source are comparable with or as great as, when compared with the wavelength, the radiation is much more complicated, so that sound is radiated with different success in different directions.

V. DOPPLER EFFECT

The so-called Doppler effect is responsible for the fact that the pitches of the ambulance sirens drop as the ambulance passes. It is hardly musically relevant, but because many people wonder about it, it might be worthwhile to present it while sound waves are under discussion.

The pitch we perceive of tones is determined by the frequency at which our eardrums vibrate, i.e., how frequently overpressures and underpressures hit them. In air, such pressure waves dash away from the source at a velocity of about 340 m/sec. If the listener approaches the source, the pressure maxima must hit his or her eardrums at a higher frequency than if the distance between them remained constant. When the listener is moving away from the sound source, the opposite must happen. We may think of riding a boat on wavy water. When the boat moves against the waves, the waves will hit the stern more frequently, and under conditions of tail wind, they will hit less frequently.

This can be expressed in more precise form. Let the velocity of the ambulance be V (m/sec) and let one of the sirens have a frequency of f (Hz). As long as the ambulance is traveling considerably slower than the speed of sound (a truly realistic assumption), the perceived pitch is determined by the frequency f' and the speed of sound according to the following equation:

$$f' = \frac{c+H}{\lambda} = f\frac{c+H}{c} = f\left(1 + \frac{H}{c}\right) \qquad (2.4)$$

which is a frequency *higher* than that of the siren. When the ambulance has passed, the distance between the listener and the ambulance increases, and in this case plus changes to minus in the equation.

Let us take a concrete example. Let the velocity of the ambulance be 75 km/hr, or 20.8 m/sec. Let the frequency of one siren be 220 Hz, and the speed of sound, 340 m/sec. When the ambulance is approaching, the perceived pitch is determined by the frequency f':

$$f' = 220\left(1 + \frac{20.8}{340}\right) = 220(1 + 0.06) = 233.2 \text{ Hz}$$

When it has passed the listener, the frequency f'' determines the pitch

$$f'' = 220\left(1 - \frac{20.8}{340}\right) = 220(1 - 0.06) = 206.8 \text{ Hz}$$

If we consult the appropriate table in Chapter 4, we will conclude that the pitch dropped by about a major second in this case.

VI. SOUND AND OBSTACLES

Our environment is full of obstacles to a free propagation of sound waves. The effect of an obstacle on sound mainly depends on the size of the obstacle as compared with the wavelength of the sound. If the obstacle is small relative to the wavelength (low pitch, small obstacle), the sound finds its way around it, and no sound shadow occurs behind the obstacle. In the opposite case (high pitch, large obstacle), the sound cannot travel around the obstacle, so that a sound shadow appears behind it. Thus, short waves cannot travel around large obstacles. Again, we can compare this with water waves. A small pole will prevent only small waves, but a pier certainly will prevent even large wakes.

In principle, high pitches cannot manage obstacles whereas low pitches do. For this reason it is often somewhat more difficult to perceive speech clearly when listening behind rather than in front of a speaker. It is mainly the low-frequency components that manage to travel around the speaker's head, and these are not sufficient for understanding speech. The poor success of high frequencies to travel around corners also contributes to the effect that we tend to perceive our own voices as darker than our listeners do: The high-frequency components in the sound of the voice which contribute to a bright voice quality cannot travel from the lip opening to the ears as successfully as the low-frequency components. Other factors contributing to this effect will be presented later in Chapter 5.

VII. SOUND INTENSITY MEASURES

Sound propagation implies that the sound waves dash away from the sound source at a speed of about 340 m/sec. If we stand in a place where a sound wave passes, we can measure pressure oscillations reflecting the passing by of pressure maxima and minima. These pressure variations have the same frequency as the sound source, provided that the distance between source and listener is constant, i.e., provided that there is no Doppler effect. In other words, if the source emits a sine wave signal at 440-Hz frequency, sinusoidal pressure variations of this frequency are generated at any point along the directions of sound propagation.

It was mentioned that the intensity we perceive of a sound depends on the amplitude of the sound pressure oscillations. This amplitude is usually symbolized by p. The unit used for measuring p is now pascal (Pa), as mentioned. One Pa corresponds to the pressure arising when the force of 0.1 kg (approximately an apple) is distributed over an area of 1 m^2. Earlier, other units were used:

$$10 \text{ dyne}/cm^2 = 0.0001 \text{ bar} = 1 \text{ millibar (mb)} = 1 \text{ newton}/m^2 1 = 1 \text{ Pa}$$

The softest perceptible sound pressure amplitude is about 0.00002 Pa and the loudest we can hear without pain is between about 20 and 200 Pa depending on the frequency.

Thus: Sound pressure amplitude, symbolized p, is given in the unit Pa. If the amplitude is given in RMS value, the sound pressure amplitude corresponds to the

VII. Sound Intensity Measures

maximum deviation during one period from the atmospheric pressure multiplied by $1\sqrt{2}$.

The sound pressure amplitude decreases with the distance to the sound source. If we want to specify how loud the sound is that a sound source emits, it is necessary to give the sound pressure amplitude at a certain distance. However, this is not always all that efficient, because sound sources may radiate very differently in different directions, and sound reflections complicate the situation further. Therefore it is better to specify the *power* of the sound source, *P* (i.e., how much work the source produces per time unit when it keeps the air particles in motion). The power is quantity of work per time unit. It is measured in the unit watt (W), earlier also in horsepower, 1 kW = 1.36 horsepower. Table 2.3 shows approximate maximum power for some different sound sources.

Thus: *The power P is measured in watts and specifies how much work the sound source is doing per second.*

The power of the sound source informs about the loudness of the sound that reaches our eardrums only in an indirect way. For this purpose it is more informative to specify what the power is *over a given area*. Power per unit area is called *intensity (I)*, and the unit is simply W/m^2.

Thus: *The intensity I is measured in watts per square meter and tells how much energy is being transported per second over an area of 1 m^2.*

We have all experienced how the sound intensity decreases with the distance to the sound source. If you hold the telephone close to your ear, you can hear quite well, but if you put it on the table in front of you, it is difficult to hear. This is no wonder. Let us first assume that the distance between the earphone and the ear is 1 cm. In that case the sound power is distributed over half a sphere with a radius of 1 cm (thus assuming that the earphone radiates no sound backward). The area *A* of that half sphere is

$$A = \tfrac{1}{2} \cdot 4\pi r^2 = \tfrac{1}{2} \cdot 4\pi \cdot 1 \cdot 1 = 6.3 \text{ cm}^2$$

Table 2.3 Approximate values of the maximum power that can be produced by some different sound sources

Sound source	Sound power (W)
Bass singer	0.03
Clarinet	0.05
Trumpet	0.3
Trombone	6
Symphony orchestra	60

If we increase the distance to 10 cm, we get the area

$$A = \tfrac{1}{2} \cdot 4\pi \cdot 10 \cdot 10 = 630 \text{ cm}^2$$

or 100 times larger. As the distance is squared in the equation, sound intensity decreases with the distance squared. To tell the truth, it actually decreases somewhat more, because some sound energy gets lost on the way, when sound is traveling in air. These losses become very important at large distances.

Now we have heard about the sound pressure p, the sound power P, and the sound intensity I. The relationships between these can be described in formulas. According to physics, power P is the product of velocity V and force F:

$$P = V \cdot F \tag{2.5}$$

The velocity concerned here is that of the air particles, not the speed of sound propagation. Further, the pressure p equals force F per area A

$$p = \frac{F}{A} \tag{2.6a}$$

Thus, pressure is proportional to force:

$$p \sim F \tag{2.6b}$$

According to Eq. (2.6b), we can substitute pressure p for the force F in Eq. (2.5), provided we change the equality sign to a proportionality symbol.

$$P \sim V \cdot p \tag{2.7a}$$

Further, the particle velocity V is proportional to pressure p:

$$V \sim p \tag{2.7b}$$

If we insert this expression in Eq. (2.7a), the result is that

$$P \sim p \cdot p = p^2 \tag{2.7c}$$

i.e., power is proportional to the pressure squared. The same applies to intensity, which is nothing more than power per unit of area:

$$I \sim p^2$$

The reader who feels uncomfortable with equations can observe that this was a very complicated way of telling the simple truth, that power and intensity are proportional to the pressure squared.

VIII. DECIBEL AND SOUND LEVEL

Under favorable conditions, we may perceive sounds already at pressure amplitudes as low as 0.00002 Pa, and if the pressure amplitude soars to 20 or 200 Pa, the sensation of ear pain appears. The range of amplitude that we can hear is thus

VIII. Decibel and Sound Level

enormous. For this reason it is impractical to use sound pressure amplitude in Pa. Instead a logarithmical measure is used, the decibel (dB). This measure relieves us not only from lots of decimals, but it also provides a measure that matches the characteristics of our hearing much better than the pressure measure. This is illustrated in Figure 2.4.

What is a logarithm? The logarithm of a number a is that number N to which a given number, which is called the *base,* should be raised to equal the number a. To an unexperienced reader this, of course, sounds very confusing, so let us take a few examples. Often 10 is chosen as the base. In any event, the logarithm for 10 is 1, or log 10 = 1, because 10 raised to the power of 1 equals 10. Similarly, log 100 = 2, because $10^2 = 100$. And log 1,000 = 3, and log 10,000 = 4. However, log 1 = 0, because $10^0 = 1$. Consequently, log 2 must be a number between 0 and 1, namely, 0.3010. Likewise, there is a logarithm for any number, but how does one get hold of them? The simplest way is to get a pocket calculator, plug in the number, and then push the button labeled "log."

This is certainly a very simple solution, but it may still be advantageous to know a bit more about what a logarithm is and how it is derived from the real number. For the sake of simplicity, we will stay with the logarithms using the number of 10 as the base.

In the logarithm, the figure(s) to the left of the decimal point equals the number of figures to the left of the decimal point in the real number. The figures to the right

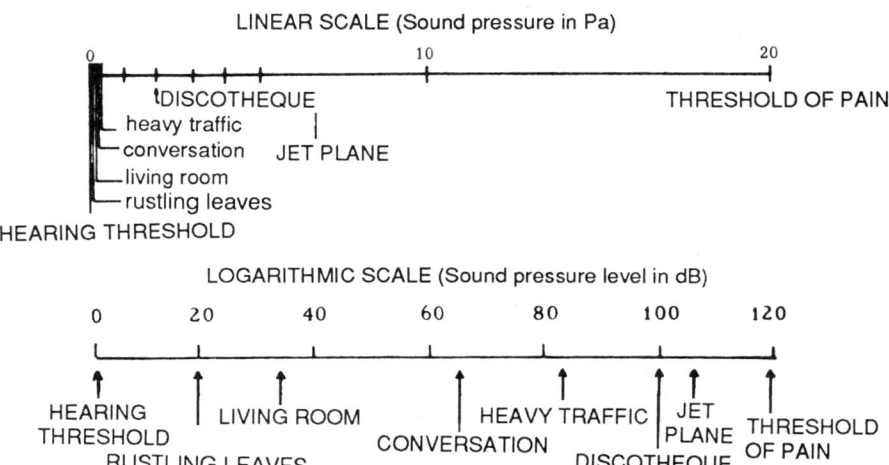

Figure 2.4 Intensities of some different sounds given in linear and logarithmic measures. (From Spens, 1970.)

of the decimal point in the logarithm is the logarithm for the *entire* number, complemented by zeros at the end, if required.

Examples:

$$\log 2 = 0.301$$

$$\log 20 = 1.301$$

$$\log 200 = 2.301$$

The logarithms were no less than a revolution to mathematics when they were invented, because multiplication and division, which are somewhat tedious processes, were transformed into simple addition and subtraction of the logarithms for the numbers concerned. Let us take an obvious example: 100 · 1,000. This task is reduced to adding the logarithms: 2+3 = 5. But what does this number 5 mean? It means that if the base is raised to the power of 5, we obtain the number we are looking for, or 100,000.

This was a simple example, so there was no problem in figuring out what number 5 was the log for. In reality, things are generally much more complicated. In such cases, the simplest solution is, again, to go to the pocket calculator, insert the log value obtained, and then press the button marked 10^x. Try this with the number we just had, 5! But try also with the log 0.301, which we met a while ago.

Conversely, division becomes subtraction of the log values, squaring becomes doubling, and taking the square root corresponds to halving the log. The slide rule, which earlier could be seen peeping out from the breast pocket of almost any engineer, was based on this fantastic simplification. However, nowadays, slide rules are rare, because they have lost the competition with the pocket calculator.

This was a small digression, so back again now to the main track. The great advantage with using logarithms in acoustics is the fact that the number of figures required is reduced: log 2 = 0.301 and log 200,000,000 = 9.301.

The logarithmic dB measure is also *comparative*. It tells how many times smaller or greater the amplitude is as compared with a reference value. The reference is often a standard, namely, the smallest amplitude that an average ear can hear at 1,000 Hz. This value corresponds to a pressure of 0.00002 Pa and an intensity of 10^{-12} W/m². Thus, the trick is to divide the pressure or intensity by these references and then calculate the log of the ratio. What is then obtained is a measure of sound intensity in a unit called *Bel*. Generally the Bel unit is a bit large, so the tenth of it is the normal unit (i.e., dB).

As mentioned, the dB measure is comparative and logarithmic. Therefore, it should be clearly distinguished from other sound intensity measures. One way of promoting this is to call the magnitude that is measured in dB *level* rather than *amplitude*. Thus, intensity level and pressure levels are measured in dB, whereas intensity and pressure are measured in watts per square meter and Pa, respectively. The intensity level L_I is 10 times the log of the ratio between the intensity I and the reference intensity I_{Ref}:

VIII. Decibel and Sound Level

$$L_I = 10 \cdot \log \frac{I}{I_{\text{Ref}}} \text{ (dB)} \tag{2.8}$$

It is not necessary to use the standard reference; any reference is fine. If one chooses a reference greater than the amplitude of the sound, the ratio will be smaller than 1. This implies that the log is smaller than 0, i.e., negative. This is a consequence of the fact that log 1 = 0, and then, the log of a number smaller than 1 must be negative. If the intensity level difference is −3 dB, the log of the intensity ratio is 0.3, which corresponds to an intensity ratio of 1/2 (those who doubt this should consult their pocket calculator).

According to Eq. (2.7d) we must square the pressure to obtain something proportional to the intensity. Squaring corresponds to a doubling of the log, as mentioned. To obtain the sound level, it is necessary to square the ratio between the pressure and the reference pressure. This means that the log of the ratio must be doubled, if we measure pressure level in Bel, or multiplied by 20, if we measure in dB. If the value of 0.00002 Pa is used for reference, the term *sound pressure level* (SPL) should be used, and if any other reference is used, the term is *sound level*. In terms of an equation, the sound level L_p for a sound of pressure amplitude p is

$$L_p = 20 \cdot \log \frac{p}{p_{\text{Ref}}} \text{ (dB)} \tag{2.9a}$$

and

$$\text{SPL} = 20 \cdot \log \frac{p}{0.00002} \text{ (dB)} \tag{2.9b}$$

The level measure decibel is fundamental in acoustics, so let us do some exercising. It is essential to remember that level measures refer to a ratio between two intensities or pressures. For example, what is the level corresponding to a doubled intensity? Log 2 = 0.301 and 10 · 0.3 = 3 dB. Similarly, the intensity ratio of 10:1 corresponds to 10 dB: log 10 = 1, and 10 · 1 = 10 dB. These two levels might be worthwhile to remember: A doubled intensity is 3 dB and a 10-fold intensity is 10 dB.

According to Eq. (2.7d), intensity is proportional to the pressure squared. This is reflected in Eq. 2.8 and 2.9: The log of the pressure ratio is multiplied by 20 but that of intensity ratios is multiplied by 10. Let us take an example. Suppose two sound intensities have the ratio 2:1. This implies that the pressures of these same sounds have the ratio of $\sqrt{2}:1$, because it is the pressures squared that are proportional to intensity. Log $\sqrt{2}$ = 0.15, and therefore the level is 20 · 0.15 = 3 dB. Thus to our great relief, we see that if the intensity level of two sounds differs by 3 dB, the pressure levels also differ by 3 dB.

Let us also get acquainted with the levels corresponding to a twofold and 10-fold pressure. Log 2 = 0.3, and 20 · 0.3 = 6 dB; log 10 = 1 and 20 · 1 = 20.

Thus, for sound pressures a doubling corresponds to 6 dB and a 10-fold pressure is 20 dB. Observe also that a doubled pressure is equivalent to a fourfold increase of intensity. Table 2.4 exemplifies the relations between various levels and their pressure and intensity ratios.

Normally the pressure or intensity of a sound is specified in terms of the level. The main effect of using the logarithmic level measure is that the difference between small pressures or intensities is expanded, whereas the difference between great pressures or intensities is reduced as can be seen in Table 2.4. This also corresponds much better to the way our sense of hearing perceives sound. The difference between the noise of rustling leaves and heavy traffic is considerable, but the difference in loudness between a discotheque and a jet plane is not so terribly great. If measured in sound pressures, the first difference is minute and the latter is enormous. If measured in levels, the situation is reversed, which is more reasonable.

We have not spoken much about power levels, and these are rarely of any great concern. However, they obey the same principles as intensity levels, because power and intensity are both measures of energy.

IX. COMPLEX TONES

When we speak about tones, we rarely mean the simple sine wave tones that were described above. All traditional music instruments produce so-called com-

Table 2.4 Examples of the relation between levels and ratios

Level (dB)	Intensity ratio	Pressure ratio
0	1.0:1	1:1
1	1.3:1	1.1:1
2	1.6:1	1.3:1
3	2.0:1	1.4:1
4	2.5:1	1.6:1
5	3.2:1	1.8:1
6	4.0:1	2.0:1
7	5.0:1	2.2:1
8	6.3:1	2.5:1
9	7.9:1	2.8:1
10	10:1	3.2:1
15	32:1	5.6:1
20	100:1	10:1
30	1,000:1	32:1
40	10,000:1	100:1

IX. Complex Tones

plex tones. This implies that an entire family or *spectrum* of tones is sounding simultaneously, almost as a chord consisting of many tones. All these sine wave components belonging to the same spectrum are called *partials*. The first partial is often referred to as the *fundamental,* and the remaining partials are then called *overtones* or *harmonics*.

A spectrum is thus built up by partials. This implies that each air particle is vibrating at several frequencies simultaneously. This may appear as completely impossible, and yet it is quite possible. One might get an idea as to how it may happen if one thinks of a conductor conducting a piece in two-fourth time while suffering from a hand tremor. His hand is then moving up and down at two frequencies simultaneously; the tremor frequency and the beat frequency of the music.

Another way of imagining the same phenomenon is shown in Figure 2.5. Let us assume that we observe how the sound pressure is varying in time. The figure shows sound pressure versus time. Each moment of time is represented by a stick, and on each of these sticks there is a straw, the length of which reflects the sound pressure amplitude at the corresponding instance. At the bottom there is a series of straws reflecting the sound pressure variations of the first partial. On a shelf above it there is another series of straws representing the sound pressure wave of the third partial, which has a frequency three times that of the first partial. If that shelf is removed, the straws resting on it will fall down on those representing the first partial. The upper contour that then results shows a sound pressure waveform, which may arise when these two partials sound simultaneously. This waveform therefore represents a complex tone consisting of these two partials. On the uppermost shelf there is another series of straws corresponding to the fifth partial. If this shelf is also removed, one obtains the sound pressure waveform shown by the top contour in the bottommost figure. This contour now shows the sound pressure waveform of another complex tone that contains these three partials, number one, three, and five. If the sound pressure has this waveform, the complex tone consists of these three partials.

It can be realized that the exact waveform depends on the phase relationships between the partials. This is clear also from the left series of panels in the figure. There the sound pressure waveform for the second partial is modeled by an iron sheet that can be slid under the straws representing the first partial. The three lower panels on the right show three different phase relationships between the first and the second partial.

Figure 2.5 thus shows how the sound pressure values of the various partials are simply added together in each instant. It also demonstrates how the appearance of the resulting complex waveform depends on the frequencies, amplitudes, and phases of the partials.

Before leaving this figure it is fair also to point out the limitations of the analogy with air pressure values. In the figure, air pressure values correspond to straw lengths. However, although straw length is always positive, air pressure values can

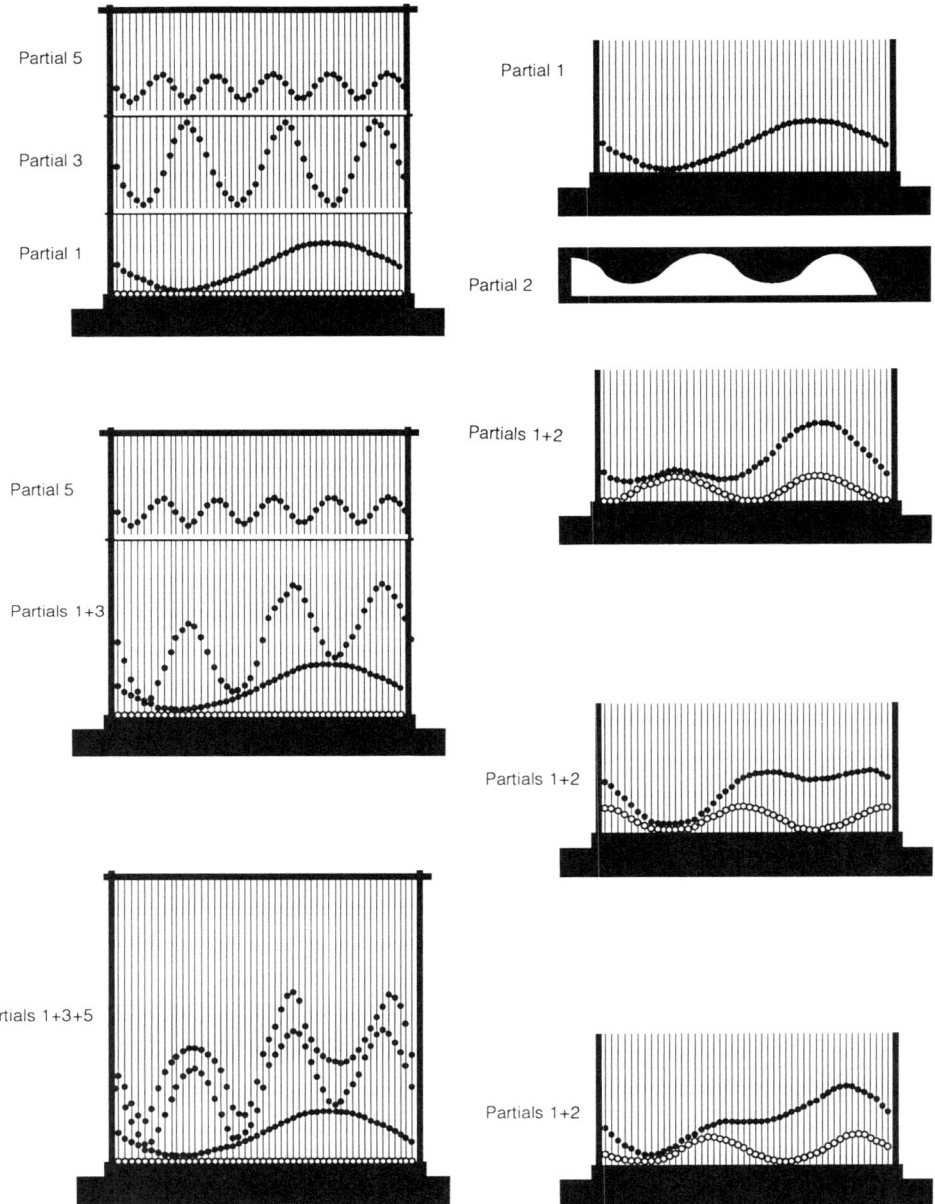

Figure 2.5 Illustration of how momentary values of the pressures for simultaneously sounding tones are linearly added when complex tones sound (see text). The *left column* of figures shows how the sound pressure waveforms corresponding to partials 1, 3, and 5 are added. The *right column* of figures shows different waveforms for partials 1 and 2 resulting from different phase relationships between these two partials. (From Taylor, 1965.)

be both positive and negative. It is important to realize that a negative pressure value subtracts from a simultaneous positive pressure value. This property is not represented in the figure.

Fourier was the name of a mathematician who demonstrated that each periodic waveform can be regarded as constituted by a series of simple sine tones with given frequencies, amplitudes, and phases. An analysis of a complex waveform that reveals the constituents of a complex tone in these terms is often called *Fourier analysis,* but perhaps even more often *spectrum analysis.*

The simple truth is that the sound pressure is oscillating at several frequencies simultaneously as soon as the sound pressure oscillates periodically but not according to the form of a sine wave. In general one can say that the more abrupt changes a waveform contains, i.e., the less similar it is to a sine wave, the more and stronger high overtones it contains.

Our hearing can sometimes single out various partials in a complex tone so that we hear them as individual pitches. This occurs, particularly if the tone does not change with time, but remains perfectly constant or stationary. The partials contribute to the timbre, and in Chapter 4 we will see that they play a very important role in chords, intervals, and scales.

As the sound levels of the partials are very significant to the sound of the signal, it is often relevant to specify which partials a complex tone contains. The phase relationships between them is of minor interest, as they normally lack significance to the sound perceived in such cases. A diagram showing the frequencies of the partials along the horizontal axis (abscissa) and their levels along the vertical axis (ordinate) is called a *spectrogram.* Often the strongest partial (which in many cases is *not* the lowest one) is arbitrarily given the level of 0 dB, and then the levels of the other partials are negative. Figure 2.6 shows an example. The spectrum envelope is of particular relevance in many cases. The envelope is a smooth curve that reflects the contour of the harmonics. We will return to this later.

Most music instruments generate waveforms in which one period of the fundamental contains integer numbers of complete periods for each of the overtones. This implies that the frequencies of the partials constitute a multiplication table, or, to use a more sophisticated term, a *harmonic series.* Such a series is characterized by the fact that all numbers belonging to the series are integer multiples of the lowest number. An example is

$$f_1 = 1 \cdot f_1,$$
$$f_2 = 2 \cdot f_1,$$
$$f_3 = 3 \cdot f_1,$$
$$f_4 = 4 \cdot f_1,$$
$$f_5 = 5 \cdot f_1,$$
$$f_6 = 6 \cdot f_1, \text{ etc.}$$

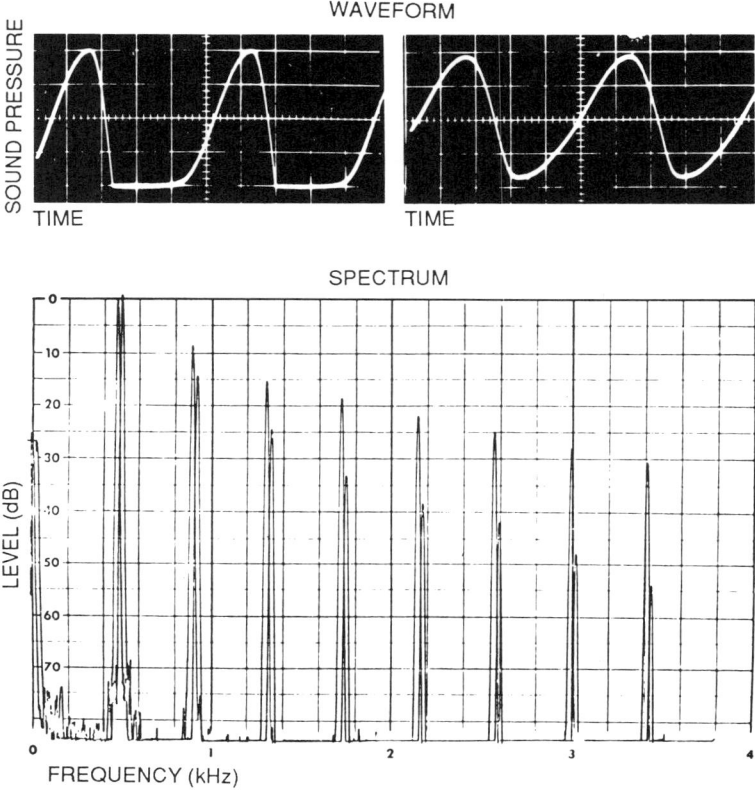

Figure 2.6 Waveform (above) and spectrum (below) of two sounds. The left waveform corresponds to the spectrum represented by the left partials in the pairs of partials shown in the spectrogram.

If we select the pitch A2* with fundamental frequency of 110 Hz, the frequencies of the overtones are 220, 330, 440, 550, 660, etc. As we will see more in Chapter 4, which deals with musical scales and intervals, we find consonant intervals between the lower partials, which in the above example are A2, A3, E3, A4, C#5, E4, etc.

Partials that have frequencies forming harmonic series are called *harmonic*. Wind instrumens produce harmonic spectra. Spectra that are not harmonic are

---

*In this book we will use a convention for naming pitches in which the octaves are simply numbered from 0 for the lowest one that gives a pitch and onward. The bottom note in each octave is the pitch C. The tone that is called "middle C" is called C4, so the pitch A4 has the frequency 440 Hz, which makes it a simple thing to remember. This convention seems far more straightforward than its alternatives.

called *inharmonic*. Percussion instruments produce such spectra. The timbre of inharmonic spectra is unstable and changes more or less constantly.

X. SEVERAL SIMULTANEOUS SPECTRA

We just saw that the sound pressure values of different tones are simply added in each moment. This, of course, is true not only when the tones originate at the same place, but also when they come from different sources.

Imagine we stand in a room where a clarinet is playing the pitch A3 (220 Hz) and an oboe is simultaneously playing the pitch E4 (330 Hz). The third partial of the clarinet has the frequency of 3 · 220 = 660 Hz. It can easily be realized that this is also the frequency of the second partial of the oboe: 2 · 330 = 660 Hz. Thus these spectra have a common partial at the frequency of 660 Hz, which corresponds to the pitch of E5. In any point in the room, the amplitude of this partial depends on the phase of this partial. This phase depends on two factors: the point's distance from the instruments and the wavelength of the tone. This is illustrated in Figure 2.7.

In certain points such as point P1 in the figure, the third clarinet partial is in phase with the second oboe partial, and in other points they are in counterphase, such as in point P2. Let us assume that the amplitudes are equal. Then, the sound pressure amplitude will double in points of phase agreement, so that the sound level will be raised by 6 dB according to what was explained previously (This chapter, Section VIII). In counterphase points there will be a complete cancellation if the amplitudes there are equal. This means that the amplitudes of the spectrum partials vary considerably in a room. This applies even when only one instrument is playing, because all points where the sound is reflected act as additional sound sources. To be more accurate, one can instead imagine a system of mirror pictures of the source, or mirror sources, located in the same places as wall mirrors would

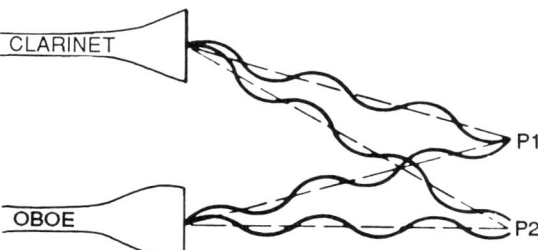

Figure 2.7 Dependence of the sound pressure on the point of observation. In point P1, sound pressure waves from the two instruments are in phase, so that the resulting amplitude is quite high. In point P2 they are in counterphase, so that the pressures cancel each other if the amplitudes are equal.

seem to place the source, i.e., outside the room. The number of such mirror sources grows rapidly if all walls, the floor, and the ceiling are good reflectors. We will see more about this in Chapter 8, which deals with room acoustics.

Many instruments radiate sound from several places simultaneously. The flute is one example. It leaks sound both from the embouchure and the first open hole or holes. For this reason the spectrum of a flute tone depends very much on the place of observation, and it is difficult to define it.

XI. BEATS

Above we have seen that the phase relations between spectrum partials are not very important for the sound perceived. However, we have also seen that the phase relation between two simultaneously sounding sine tones is quite decisive for the amplitude. Here we will see another case, called *beats*.

Beats arise when two sine tones sound simultaneously that differ slightly in frequency. Let us take a concrete example. Tone 1 has the frequency of 100 Hz, and tone 2, 102 Hz. Tone 1 thus attempts to raise and lower the sound pressure 100 times per second, and tone 2 tries to do the same thing 102 times per second. Then, the phase between these two tones must vary. Two times per second they will be in phase, and two times, they will be in counterphase. If we then take into account that the sound pressure simply is the sum of these two pressures, we realize that the amplitude will vary regularly. Two times per second it will reach a maximum and a minimum, as is illustrated in Figure 2.8. It is such regular amplitude variations that are called beats. In any event we can conclude that two sine tones with almost the same frequency and similar amplitudes will beat at a rate equal to the frequency difference.

Earlier, when there were no frequency meters available, beats were used to measure frequency. Sets of tuning pipes with exact and known frequencies served as standards. A pipe with a frequency almost coinciding with the one to be measured was selected. Then, it was just a matter of counting the beats while looking at the watch.

Also tones that form slightly mistuned consonant musical intervals beat. The reason for this is that the tones normally have harmonic spectra, so they contain harmonic partials. Let us again take a concrete example. Two tones with the frequencies of 220 and 330 Hz constitute a pure fifth. But as we just saw, the second partial of the higher tone and the third partial of the lower tone both have a partial at 660 Hz. If one of the tones is a bit sharp, say, 331 rather than 330 Hz, its second partial will be at 662 Hz, so that it will beat 2 times per second with the other tone's third partial at 660 Hz.

As a matter of fact, beats occur even in the case of slightly mistuned consonants between pure sine waves. The beats are quite weak in such cases, but still they are

XII. Resonance

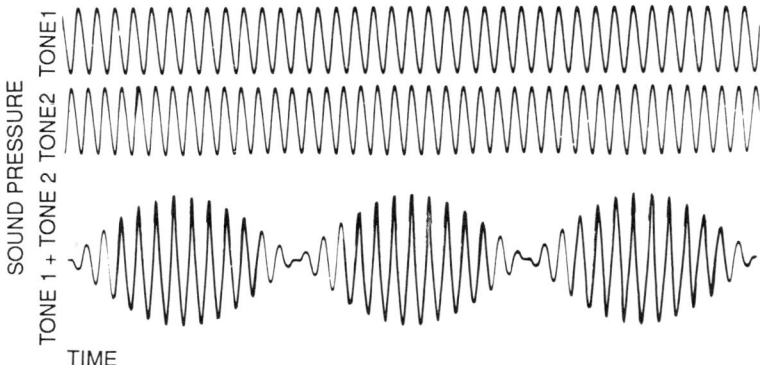

Figure 2.8 Example demonstrating why two tones of nearly the same frequency give rise to beats if sounding simultaneously. When the sound pressures are in phase the summed amplitude becomes large, and when they are in counterphase the resulting amplitude is reduced to zero if the two tones are equal in amplitude. The frequency of occurrence of phase and counterphase relationship, i.e., the beat frequency, equals the frequency difference between the tones.

often clearly audible. They arise because overtones are being generated in our ears. We will return to this phenomenon in Chapter 3 dealing with hearing.

The beats between nearly just intervals played with complex tones with harmonic spectra have been quite important in the development of scales and tuning systems. We will look more into this in Chapter 4, dealing with scales.

Beats correspond to a regularly undulating sound pressure amplitude, which is typically the result of a regularly varying phase relationship between two tones. Beats are sometimes confused with vibrato, although this is a quite different phenomenon. Vibrato is typically used in wind and bowed instruments. Vibrato is different from beats in the sense that it corresponds either to a regularly varying fundamental frequency or to a regularly varying sound level.

XII. RESONANCE

Sound waves are reflected when they hit a change in the density of the medium. Such a change may be caused by a barrier, such as a lid at the end of a tube, but also in fact, by an aperture in a tube. Reflection implies that the direction of sound propagation is changed. The principle is the same as with reflection of light: The angle of incidence is identical with the reflection angle. Thus, if the reflecting surface is normal to the direction of sound propagation, the direction of propagation is reversed: The sound wave returns. Reflection is the origin of an acoustic phenomenon of immense importance, namely, resonance, which occurs in resonators.

All systems that possess both mass and compliance and in which there is reflection serve as a resonator. *Mass property* means possession of weight, a condition that is almost universally met. The mass property implies that it is a bit difficult to change the speed. We all know that it is difficult to stop a heavy object in motion. *Compliance* means that the object strives to resume its original volume if it has been compressed. We just saw that reflection means switching of direction of propagation.

A car is an extreme example of a resonator. It is heavy, so it has a considerable mass. Because of the springs or bumpers, it returns to its original altitude above the street, if it is pressed downward. The air column within a tube is another example of a resonator. Air is light but not void of mass. If air is compressed, it strives to resume its original volume; just think of how it feels to use a bike pump. Probably the most frequent example of air column resonator is the human vocal tract, the tube constituted by the mouth and the pharynx. Rooms can be regarded as huge resonators.

Resonance is manifested such that the resonator favors oscillations at certain frequencies, the resonance frequencies. Thus if one hits a resonator, it starts oscillating on these resonance frequencies. If one strikes one end of a tube open in both ends against the palm, one can generally hear a quickly decaying tone, corresponding to one of the resonances. If one sends sound through a resonator with prominent resonance frequencies in the vicinity of some 100 Hz, one would probably call the sound "hollow." If one tries to bump a car up and down, one notices that it cooperates only at a certain bumping frequency, its resonance frequency: It certainly does not pay to deviate from that frequency if one wants to generate high amplitudes. If one varies the frequency of a sine wave tone sounding in a resonator, one finds that the amplitude becomes much greater at certain frequencies, the resonance frequencies. If one tries to send an entire spectrum through a resonator, those partials that are closest to the resonance frequencies will reach a considerably higher amplitude than the others. This, by the way, is the basic principle of vowel formation in the human voice, as we will see later.

It is not difficult to make oneself a quite concrete idea of how resonance occurs. Imagine that you are in one end of a long tunnel closed at both ends. If you clap yours hands, the sound of the clap will travel to the opposite end, where it is reflected. After a little while it has returned to the place where it originally was created. If in that very moment, when the clap sound returns, you again clap your hands a second time, the sound must obviously be amplified. The effect will increase still more, if you repeat the trick, clapping your hands again when the already amplified clap sound returns. One can use this effect systematically by clapping one's hands at regular intervals, i.e., at a certain frequency, which have been adapted to the travel time of the sound through the tunnel or, in other words, to the travel distance.

This is exactly the situation in a resonator oscillating at resonance: Direct and reflected sound cooperate so that the amplitude is increased. The amplitude of the resulting sound merely depends on the success with which the sound has managed to keep its amplitude during the travel, forth and back.

After some thinking one can realize that there are several frequencies at which direct and reflected sound cooperate to increase the resulting amplitude. One can send away two hand clap sounds, i.e., double the frequency and still keep synchrony between direct and reflected sound. The same applies for multiplying the frequency by other factors, three, four, etc.

One can also imagine that a similar situation would arise if a loudspeaker is moved back and forth at the end of a tube. It causes rarefications and condensations of air particles, and these pressure oscillations dash away to the opposite end of the tube where they are reflected. The magnitudes of these rarefications and condensations must depend on the phase relation between the motions of the cone and the reflected sound wave. And the travel time of the sound in the tube decides the resonance frequency.

Resonance occurs everywhere in our environment, and it has a decisive effect on the sound generation and pitch control in music instruments. Tubes act as resonators even if one or both ends are open or closed. Strings are other examples of resonators; they possess both mass and compliance, and motion is efficiently reflected at the string supports. We will look more closely into various types of resonators in later chapters.

XIII. ATTENUATION

The magnitude of the amplitude at resonance depends on the amount of attenuation contained in the tube. Attenuation is caused by losses, which occur for various reasons. Some oscillation energy is converted to heat, and in a tube resonator some is lost when the air particles scratch the tube walls. Sound energy may also be lost at the reflection; for instance, sound is lost to the outer air when it is reflected at the open end of a tube. One can also increase the attenuation by inserting porous material in the tube. This effect can easily be experienced. Sustain the vowel /i/ (as in the word *heat*) loudly and while doing this, put a cotton ball up to your lip opening. The timbre of your voice is then clearly affected; perhaps you would like to call the sound quality muffled. The effect arises because the attenuation of your vocal tract resonances is increased by the cotton, trapping sound energy. Next time, repeat the experiment, but now bring a hard object, such as a tablespoon, up to your lip opening. The spoon does not appreciably change the attenuation, and the sound is changed in a quite different way.

If the resonator has small losses, i.e., low attenuation, the oscillations in it decay slowly, and the sound at the resonance frequencies is much louder than at nearby frequencies. An example of a resonator with low attenuation is the piano string. The low attenuation contributes substantially to the slow decay of the piano tone. A violin string is more attenuated; the decay in a pizzicato is much quicker. The main difference between these two types of strings depends on the reflection at the ends, which is much more efficient in the case of the piano. A fingertip is soft.

The attenuation at the resonance frequencies can be measured in two ways. One is to observe how quickly a tone corresponding to a resonance frequency decays after the excitation has been stopped. Another alternative is to measure how large frequency change is needed on both sides of the resonance to reduce the sound level in the resonator by 3 dB. The distance between these two frequencies is called the *bandwidth* (see Figure 2.9). If the attenutation in the resonator is small, the bandwidth is narrow.

The decay of a sine tone in a resonator follows the same pattern in all resonators. If the initial amplitude is called A_0 and the amplitude at time t is A_t,

$$A_t = A_0 \cdot e^{\delta t} \qquad (2.10)$$

where δ is a constant, the so-called attenuation constant. This constant is related to the bandwidth B according to a simple formula:

$$\delta = -\pi B \qquad (2.11)$$

This means that $A_t = A_0 e^{-\pi B t} = \dfrac{A_0}{e^{\pi B t}}$ \hfill (2.12)

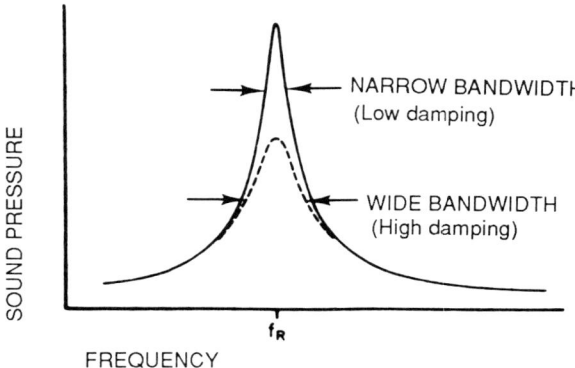

Figure 2.9 Resonance curves for two resonators with considerably differing attenuations but identical resonance frequency f_R.

XIII. Attenuation

When looking at this reasonably complex equation, it might be appropriate to inform the reader that there are two types of equation readers. One type takes the time required to understand the message of the formula. The other reads equations and the rest of the text at the same speed. The latter type tends to develop a phobia for the equal sign rather quickly. This is neither good nor appropriate. Let us therefore spend some time contemplating this equation.

An efficient way of grasping the information hidden in an equation is to assume different values for the variables and see how that affects the end result. A_0 is a constant, the initial amplitude. For $t = 0$, i.e., before the decay has started, the denominator will be $e^0 = 1$, that is, A_t at $t = 0$ equals A_0. When t is small the denominator is close to 1, so the difference between A_t and A_0 will be small, but when t grows, the difference will increase. The shape of the curve can easily be generated by inserting a few values of t and observing the end result. To do this, we need to assume a value for the bandwidth B. Start with, for example, 10. Then explore the effect of the bandwidth on the onset pattern by doubling the bandwidth, i.e., increasing the attenuation. In this way one can explore the message compressed in formulas. The time needed for this exploration is strongly dependent on training. Let us now continue by looking at the onset pattern.

At the onset of oscillation, the amplitude is built up according to the mirror image of the decay:

$$A_t = A_0(1 - e^{-\pi Bt}) \qquad (2.13)$$

Figure 2.10 shows these onset and decay patterns in a graphical form. At the resonance frequencies, a resonator with low attenuation is thus characterized by a slow buildup of amplitude, a high amplitude, and a slow decay. Conversely, a heavily attenuated resonator starts and stops quickly, and the amplitude at resonance is not all that dramatically exceeding that occurring at neighboring frequencies.

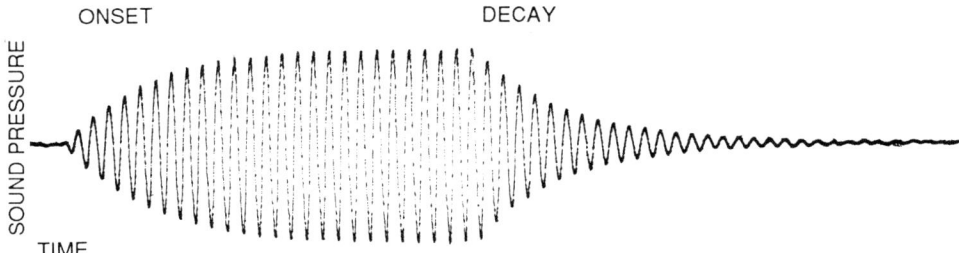

Figure 2.10 Onset and decay patterns for a resonance frequency of a resonator.

XIV. STANDING WAVES

Reflection gives rise to a peculiar type of wave propagation within a resonator called *standing waves*. Where sound waves pass freely, e.g., outdoors, the sound pressure varies between maximum and minimum, as the various parts of the sound wave passes. Thus, if we measure the amplitude in one point in the sound field, it is the same as that at nearby points if we disregard the fact that the amplitude decreases with the distance to the sound source.

Things are different in a resonator. At the resonance frequencies, stationary patterns of sound pressure amplitude occur. The reason is that the direct and the reflected sound waves cooperate at certain places and counteract each other at other places. Figure 2.11 illustrates the situation. At certain places in the resonator, the amplitude is zero; such parts of the standing wave are called the *nodes*. At

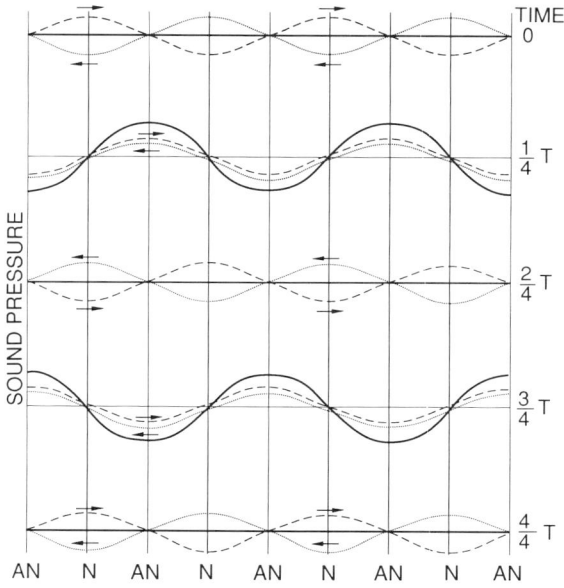

Figure 2.11 Illustration of standing waves. The graphs show the distribution of sound pressure within an enclosure, such as a tube closed by a sound source generating a sinewave at the left end (dashed curve) and by an absorption-free reflecting wall at the right end (dotted curve); in each graph arrows show direction of sound propagation and heavy curves show the resulting sound pressure. At the moment represented by the top graph the summed sound pressure is zero along the entire distance of travel. After a quarter of the period time (second graph from top) the direct and reflected sound waves are in phase along the entire travel distance, so the summed amplitude soars. After another quarter of a period the sum returns to zero and so on.

XIV. Standing Waves

certain intermediate places, the sound pressure amplitude reaches maximum; these parts are called *antinodes*. In the antinodes direct and reflected sound energy cooperate, i.e., they are in phase. In the nodes, conditions are reversed. By looking at the figure, we realize that the distance between two adjacent nodes or antinodes is only half a wavelength, so the distance between a node and the adjacent antinode is a quarter of a wavelength.

Rooms are resonators of huge dimensions. This implies that the resonance frequencies are so close to each other in frequency that it is hard to identify them. In fact, there is almost always at least one resonance at any frequency one might choose. For this reason it is possible to experience the uneven distribution of sound pressure amplitude at almost any frequency. If there is a motor emitting a low-frequency tone outside the window, it often sets up a standing wave in the room. If so, one can walk in and out of nodes and antinodes; in the nodes, no tone can be heard, but in the antinodes, the tone is very loud.

Low frequencies have long waves and high frequencies have short waves. If we let a high-frequency tone sound, e.g., by asking a friend to whistle a tone with constant pitch, it is generally enough to move the head a bit to notice a clear difference in loudness. One then moves the head between nodes and antinodes of the standing wave. If one occludes one ear, the effect is enhanced, because the possibility is then eliminated that one ear is in a node and the other is in an antinode.

When we listen to speech or music, we rarely notice the standing wave patterns. First, for higher frequencies it is rare that both ears are located in a node or antinode simultaneously, so the effect is generally reduced. Second, we rarely have any use for the information represented by the antinodes and nodes of the standing waves, and we rarely pay attention to things that do not offer us any useful information. If one listens to old monophonic recordings, which were made with only one microphone, standing waves can be quite disturbing, e.g., in terms of a consistently reappearing, roaring bass tone.

However, standing waves are no less than a disaster if one wants to make a spectrum analysis of a complex tone. It has been demonstrated that the sound level values collected at different locations in a reverberant room from a sine wave tone all differ from one another, and about 70% of these values fall within a range that is no less than 11 dB wide! Imagine what kind of messy data would result from a spectrum analysis in which each single partial has this variability. For this reason, a spectrum analysis made in a reverberant room must be regarded as wasted time. Such measurements must be made in anechoic chambers void of sound reflection and thus free from standing waves.

CHAPTER 3

Ear and Hearing

I. INTRODUCTION

The oscillations around the atmospheric pressure described in the previous chapter alternately presses the eardrum inward and sucks it outward. In this way the eardrum is forced to vibrate in synchrony with the sound pressure oscillations. As we will soon explain, these vibrations are transferred to the inner ear, where they are transformed to nerve impulses, and these, in turn, are fed to the brain. In a way, the ear can be regarded as a kind of microphone, as it transforms sound to a type of electrical signal.

The properties of this innate human "microphone," catching music in its acoustic shape, is obviously significant if we want to understand music in a more profound sense. Music sounds must depend to a considerable extent on the function of our ear and sense of hearing. Hearing must also be an important factor when the composer writes his or her music and when the musician plays it. For instance, it would hardly be rewarding for either of them to work with effects that the human ear cannot perceive. However, it must be most rewarding to tailor the acoustical form of music so that it matches the characteristics of hearing.

From this we should not conclude that music perception can be understood as soon as we understand the properties of the hearing system. When the sound waves have been converted to nerve signals, they disappear in the gray mystics of the brain, scattering in a multitude of directions. The result is the experience of music. To a large extent, the principles of these processes are poorly understood or even completely unknown. Psychology has lately developed new methods, so that it now more efficiently than before can attack the extremely complex problems that we call music experience.

In this chapter we will describe characteristics connected with the peripheral parts of the hearing system, i.e., what happens in the ear and its neighborhood. In Chapter 10 we will return to the more complex aspects of music listening, e.g., how the tones in a melodic sequence in a sense "stick together," forming intervals, motives, phrases, melodies, etc. This is clearly the interface between music acoustics and music psychology, and those who want to learn more about these aspects are referred to textbooks on psychology of hearing and music psychology.

II. ANATOMY OF THE EAR

As illustrated in Figure 3.1, the ear can be divided into three parts: the outer ear up to the eardrum, the middle ear between eardrum and cochlea, and the inner ear, or the cochlea.

A. Outer Ear

The purpose of the outer ear, or the pinna, is to catch the sound waves. The outer ears of humans are not much to brag of. Other creatures have much larger screens that they can turn in various directions. The signals hitting the pinna propagates along the ear canal and reach the eardrum. The ear canal protects the eardrum from outer damages and provides a rather constant temperature and moisture for the sensitive eardrum.

The ear canal is about 2.5 cm long in adults. It is slightly narrower at the eardrum than at the pinna. Acoustically it acts as a resonator, with a strongly peaked resonance near 4 kHz. This resonance adds to the sensitivity of the ear in this frequency range.

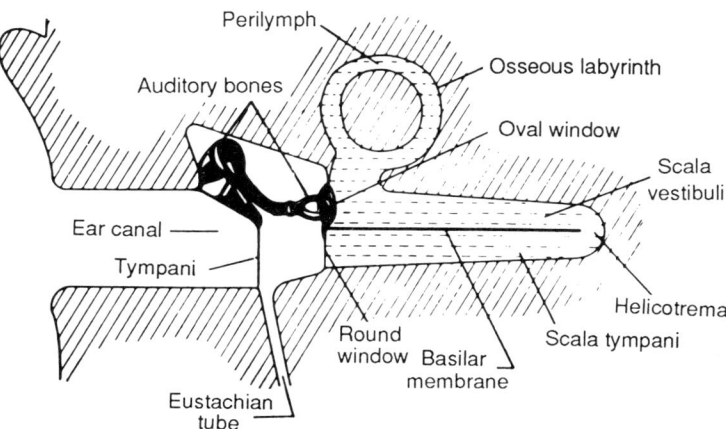

Figure 3.1 Schematic illustration of the three main parts of the ear. In reality the cochlea is coiled. (From Hadding and Pettersson, 1972.)

B. Middle Ear

In the middle ear the sound pressure oscillations are converted to mechanical vibrations. This happens at the eardrum, which for this reason can be compared with the membrane in a microphone. The eardrum is not flat but somewhat conical and pointing inward. It has an oval shape, and the area is about 0.7 cm^2. The thickness is about 0.4 mm. The mechanical vibrations of the eardrum reflect the sound pressure oscillations.

Our hearing normally has its greatest sensitivity near 4 kHz. In this frequency region, sound can be perceived when the vibration amplitude of the eardrum is no greater than 10^{-9} cm, which is slightly smaller than the diameter of a hydrogen molecule. At these small amplitudes, the vibration amplitude of the eardrum is similar to that of the air particles. At greater sound pressures, the eardrum amplitude is reduced as compared with that of the air molecules. Engineers call this *compression,* and in the ear it serves the purpose of expanding the audible range of sound pressures, because even sound of very small amplitudes can be perceived.

This compression mentioned is due to a muscle reflex that is triggered by loud sounds. It also is triggered by one's own voice: We make ourself a bit deaf for what we are telling! For sudden loud sounds, as a gunshot, the reflex is not quick enough, so amplitudes of detrimental magnitudes may reach the inner ear. As reflexes generally tend to be slower after alcohol consumption, we may see this reflex protection of the hearing system as an indication for sober concert habits!

The eardrum is connected to the cochlea in the middle ear by means of a series of small bones, small as gravel. They are called the *hammer,* the *anvil,* and the *stapes.* The shaft of the hammer is fastened to the eardrum. Its vibrations are transferred to the stapes via the anvil, and the footplate of the stapes is connected to the oval window, a small membrane serving as the entrance to the cochlea.

In the transfer of vibrations from the eardrum to the oval window, two effects are involved. First, the bones act as a series of levers. The lever effect is greatest for soft sounds, for which they may triple the force. Second, the transfer is arranged such that a minor pressure over a large area is transformed into a greater force over a smaller area. This is because the area of the eardrum is between 15 and 30 times larger than that of the oval window. This effect is identical with the one responsible for the deep impressions in floors from stiletto heels, whereas those from sandal heels are negligible. Because of this effect the transfer from the eardrum to the oval window can be amplified by a factor of almost 100 in extreme cases. This amplification is recruited by the hearing system only for the softest sounds. Thus also in the middle ear we find mechanisms that help to expand the range of amplitudes that we can perceive.

The middle ear is filled with air that is connected to the free air via a narrow channel ending in the nasopharynx. This channel is called the *Eustachian tube.* It opens during swallowing and when one opens one's mouth forcefully. A stationary pressure difference across the eardrum occurs when the outer air pressure is changed, e.g., in airplanes during takeoff and landing. Then the eardrum is slightly deformed, so that it is expanded outward or inward. This is perceived in a very particular way, somewhat as if someone was covering one's ears with the palms. When one swallows or opens the mouth widely, the Eustachian tube is opened and the air pressure difference is eliminated so that hearing returns to normal. In connection with colds one may experience some difficulty in opening this tube, and this may cause airplane travellers some discomfort during takeoff and landing. However, the problem is easily resolved by nose drops, which the air hostesses generally can offer to all those who need them.

C. Inner Ear

The cochlea transforms the mechanical vibrations in the middle ear to nerve impulses. The cochlea has two windows facing the middle ear. We just mentioned one of these, the *oval window*. The other one is the *round window*. Connected to the cochlea are also the semicircular canals constituting the vestibular apparatus. However, these are not involved in hearing; they belong to our sense of balance.

The cochlea is coiled, making about 2¾ turns. Its length is about 3.5 cm. The cochlea is split into two channels: *scala vestibuli,* which originates at the oval window, and *scala tympani,* which originates at the round window. These channels are connected at the top of the cochlea by a narrow opening, called *helicotrema* (see Figure 3.1). These channels are not filled with air, as the middle ear, but with a very viscous fluid. The structure separating the channels consists of two thin walls, as the section in Figure 3.2 shows. The lower one is called the *basilar membrane,* which faces the scala tympani. The upper one is called the *vestibular membrane,* or Reissner's membrane, facing scala vestibuli. Between these two membranes, a third narrow channel is formed, which is also filled with a liquid. The basilar membrane is the base for the inner and outer hair cells, or the *organ of Corti*. The free end almost touches a bone structure, the *tectoris membrane*. It is from the organ of Corti that the fibers of the auditory nerve originate.

If the stapes slowly increases its pressure against the oval window, e.g., because of a change in the ambient pressure, the fluid passes from the scala vestibuli to the scala tympani via the helicotrema and thus causes the round window to bulge slightly outward. This causes no hearing sensation. When the foot plate moves out and in at a faster rate, as when sound hits the ear, the fluid does not have time enough to glide through the narrow helicotrema. Instead a traveling deformation of the basilar membrane results in the form of a so-called *traveling wave.* A traveling wave is similar to the wave motion on a water surface or a rope, the end of which is quickly moved up and down. The amplitude of this traveling wave reaches a maximum at a certain place along the basilar membrane, and the location

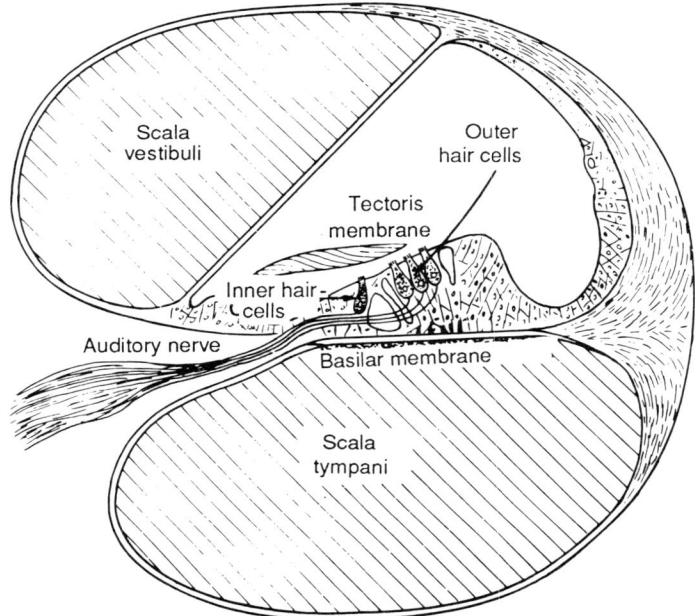

Figure 3.2 Section showing the channel system in the cochlea. (From Olson, 1968.)

of this place depends on the frequency. As we will see later in this chapter, this fact has a quite decisive significance for how we perceive sound in general and music in particular. At the place of the maximum amplitude of the traveling wave, the organ of Corti is lifted high enough to hit the tectoris membrane, which then bends these hair cells. This is what causes a hearing sensation.

The location of the traveling wave's maximum amplitude depends on the frequency, as mentioned. Figure 3.3 illustrates this. At high frequencies the amplitude maximum is located near the entrance at the oval window and the stapes. At low frequencies it is located near the opposite end of the scala vestibuli, near the helicotrema. Thus, different parts of the basilar membrane and hence different hair cells are stimulated by different frequencies. If the ear catches a complex tone consisting of many partials, several different groups of hair cells are being excited. This means nothing less than that the basilar membrane performs a spectrum analysis of the sound in the sense that it determines amplitudes and frequencies of the various tones constituting the complex sound.

Sound does not necessarily have to enter the hearing system through the main entrance of the outer ear. Sound can enter the ear also by vibrations of the skull. Thus, we hear sounds also under those conditions. This sound path is called *bone conduction*. Often it is quite useful. For instance, it allows us to hear more clearly

II. Anatomy of the Ear 43

the sound of the tuning fork if we press its end against the bone structure near the ear.

Bone conduction is also useful if one wants to investigate mechanical vibrations in objects; it is enough to press one end of a rod against the object and the other end against the bone near the ear.

Bone conduction is also relevant to the way in which one perceives one's own voice; the voice generates extremely high sound pressures within the mouth and these set up vibrations in the skull. We will return to this in Chapter 5 when speaking about the human voice.

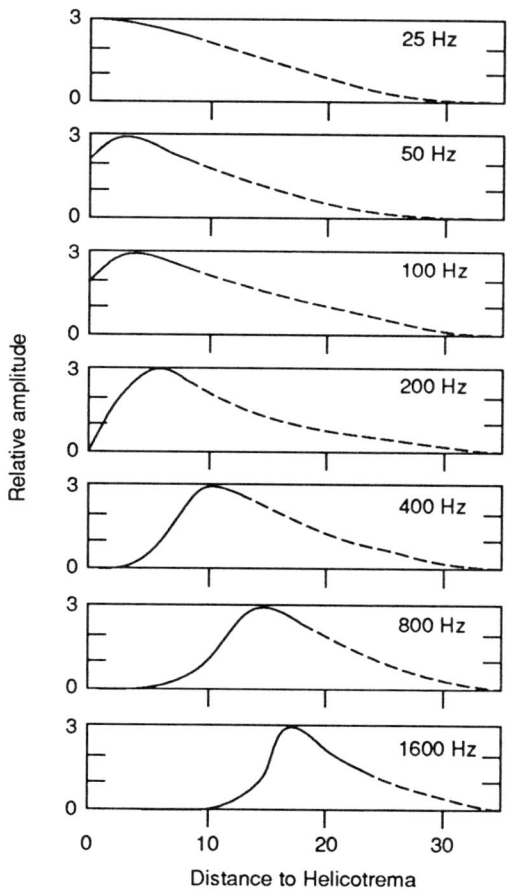

Figure 3.3 Distribution of vibration amplitude along the basilar membrane at the excitation frequencies shown.

III. HEARING

In the preceding chapter we saw how sound can be described in acoustic terms. Here we will see how we perceive these various acoustic characteristics; some of the major properties of the hearing system will be discussed.

An acoustic description is completely useless as a description of how the sound really is perceived. Our hearing system does not work as an ordinary electronic frequency, level, or spectrum analyzer, which always shows the same readings for the same signal properties (i.e., as long as the equipment is working properly!). For example, our hearing system may assure us that two tones have different pitches even though they have exactly the same frequency. This is probably the reason why many technically oriented people regard perception as subjective. Perception follows its own laws, although these may sometimes be somewhat mysterious. Therefore, in a sense, what we perceive may be as concrete and objective as what we measure. To get a good idea of acoustic properties, it is necessary to know how changes in frequency, amplitude, and spectrum sound. This relation between acoustically measured and perceived properties of sounds is the theme of the following sections.

Our experience of the environment is rich and varied, and a perceived sound contains many different aspects. The percept of a tone contains pitch, loudness, and timbre. Such aspects of a sound percept are called *perceptual qualities.* When we hear a familiar sound, we can often guess how the sound was generated. For instance we can tell from listening to the sound that it was a clarinet playing or two hard objects that clashed. In such cases we do not recognize a single pitch or a typical timbre. Rather we recognize an entire ensemble or pattern of qualities.

This part of acoustics is often called *psychoacoustics.* The author finds this term very unattractive: *Psycho* suggests a connection to the soul, and this appears truly farfetched. The topic is sometimes also called *subjective acoustics,* as mentioned, an equally repelling term, as it suggests that the information delivered by our hearing system is always somewhat unreliable and contains unpredictable components, i.e., that objective listening is impossible. A nicer term would be *auditory acoustics,* a term that, unfortunately, nobody (not even the author!) uses.

A. Pitch and Frequency

1. The Audible Frequency Range

Pitch is a perceived property, which closely corresponds to the physical concept of frequency. Thus, pitch is mainly determined by the frequency: the higher the frequency, the higher the pitch. We mentioned earlier that a very sensitive person can perceive sound that falls in the frequency range of 20–20,000 Hz. However, the audible frequency range varies with age and is considerably narrower at high

III. Hearing

ages. The frequency range that is useful for musical scales is just a part of the audible range, approximately between 30 and 4,000 Hz.

2. Pitch and Intensity

For sine tones in particular, the pitch does not depend exclusively on the frequency; the intensity is also relevant. The influence of intensity on the pitch of sine tones is illustrated in Figure 3.4 for one subject. The figure shows how intensity changes affect the perceived pitch at different frequencies. At first glance, the diagram is quite easy to read: Low frequencies drop and high frequencies rise in pitch when they get louder. However, if one examines the graph a bit closer, it is not all that easy to understand. It shows how the frequency of the tone must

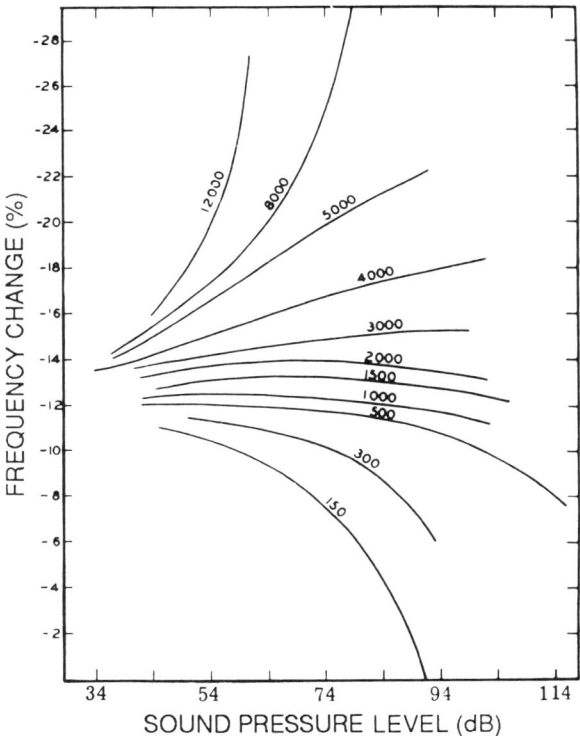

Figure 3.4 Dependence of pitch on intensity for sine tones for a subject. Graph shows frequency changes needed for keeping the perceived pitch constant when intensity is changed at various frequencies. Note that the scale is such that a falling curve implies that the frequency must be increased to retain the pitch, i.e., the pitch would otherwise drop. (From Stevens and Davis, 1938.)

be changed to keep its pitch when the intensity is changed. However, the quantity of frequency change is plotted along a negative scale, which grows more *negative* upward in the diagram. A falling curve indicates that the value grows less negative, i.e., more positive: It simply increases! If we manage to keep this in our heads, we realize that the frequency of a low-frequency note increasing in loudness has to be raised to remain constant in pitch. The reverse is true for high-frequency tones. Thus, low is lowered and high is raised when intensity is increased.

Rather substantial frequency changes are required to keep pitch constant for the subject represented in the figure. A 150-Hz tone increasing from 45 to 90 dB drops in pitch to an extent corresponding to a 12% frequency shift. This is close to two semitones in the diatonic scale. The sensitivity to this effect varies considerably between individuals. In most people it is only a fraction of what is shown in the figure.

We just saw that the tone from a tuning fork becomes much louder if bone conduction is used, i.e., if the shaft is pressed against the bone structures near the ear rather than held in the air outside the ear. The difference in loudness resulting from this is great enough to produce a clearly audible pitch shift in many subjects; the pitch is somewhat lower when bone conduction is used. However, the effect is generally rather small.

A funny consequence of this effect is that a soft sine tone at 300 Hz may sound as a pure octave of a loud sine tone at 168 Hz. The mathematically pure octave, however, has the frequency of 150 Hz. The tone that sounds as a pure octave is 12% too high. This means that mathematically it is a minor seventh! This is a good argument for avoiding confusion of perceptual and physical entities.

These relations between pitch and frequency are true for sine tones. One can easily imagine that things become much more complicated for complex tones; if such a tone contains partials in the range between, say, 200 Hz and 5,000 Hz, the pitches of the various partials depend on their intensities in different ways. In practice, the result is that the amplitude does not affect the pitch to any great extent for complex tones. Perhaps this is a good reason why complex tones are found so frequently in music. Apart from the fact that sine tones are almost impossible to generate by means of traditionally used instruments, complex tones are stable in pitch when the amplitude is changed. In certain cases the amplitude dependence of the pitch of complex tones is the opposite of that shown in Figure 3.4. If the loudness of a complex tone of about 100-Hz fundamental frequency is increased, its pitch may rise rather than drop.

3. Pitch and Pitch!

In music, pitch differences are generally given in a special unit, the semitone. These are sometimes divided into hundreths called *cents*. We will return to this in Chapter 4, dealing with scales and temperaments.

If one analyzes our perception in more detail, one finds evidence for the

III. Hearing

existence of a different kind of pitch, which does not lend itself to the semitone measure. When we listen to two persons speaking with each other, we may perhaps notice that their voices differ with regard to pitch, but still it appears farfetched to express this pitch difference in terms of musical intervals and say that Mr. Brown's voice is four semitones higher than that of Mr. Jones, on the average. And if we hear somebody pronouncing the *s*-sounds as a whistle tone with a reasonably constant pitch, it would seem awkward to describe the pitch of this whistle tone as so-and-so many semitones above some reference. Musical pitch seems alien to this kind of pitch. In other words, there seem to be two different kinds of pitch: One is used when one places tones into a musical scale, and another one is used when such a placing does not seem adequate. Thus there seem to be reasons to distinguish between pitch and pitch. Let us next view this question from a different angle.

Psychology has found that different properties of what we perceive can be quantified. For example, our perception of distance can easily be quantified: "Twice as far" means approximately the same distance for most people. It has been found that something similar applies also to other perceptual qualities, including pitch. Thus, it is not meaningless to say that one pitch sounds twice as high as another pitch. In any event, most people react in a reasonably similar way if they are asked to adjust the frequency of a tone such that it sounds twice as high as the pitch of a reference tone. This may appear as a completely mad idea to many musically oriented listeners. Nevertheless, it seems to be a fact that we have to accept.

Systematical series of experiments have been carried out in which listeners have been asked to adjust the frequency of a variable tone so that it sounds "twice as high" or "half as high" as a reference tone. By changing the frequency of the reference, the curve shown in Figure 3.5 has been obtained. This curve relates this kind of pitch to frequency. Disregarding the fact that this kind of pitch does not seem to have anything in common with the pitch, which enables us to place a tone in a musical scale, and that the relation seems to hold only for single sine tones, which are extremely rare in music, the unit used for this kind of pitch has been called *mel,* a derivative of melody! The mel scale is constructed such that a halving of the number of mels corresponds to a halving of the pitch perceived. As shown in the figure, a tone with the pitch of 1,000 mel sounds twice as high as another tone with the pitch of 500 mel. Examination of the figure tells us that this corresponds to a frequency shift from approximately 1,000 to 380 Hz.

As a curious fact, Figure 3.6 shows the magnitudes of the mel pitch differences contained in two musical intervals, octave and fifth, presented in different frequency ranges. An octave contains a small pitch difference in the bass and a great pitch difference in the treble. The pitch that is measured in mel is obviously something quite different from what musically oriented listeners mean by pitch. Because of this it is desirable or, indeed, mandatory to separate these two types of pitch. Music listeners would agree that pitch is that tone quality that allows us to place a tone in a musical scale. Henceforth, the term *pitch* or *musical pitch* will

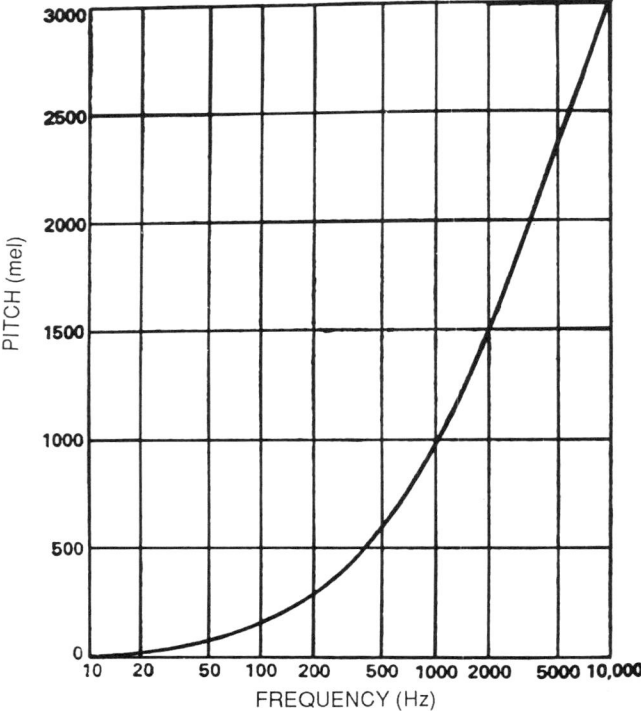

Figure 3.5 Relation between the pitch as measured in mel and frequency. (From Denes and Pinson, 1975.)

be reserved for this type of pitch. For the other type of pitch, which can be appropriately measured in mels, the term *tone height* will be used. This is the kind of pitch that comes into one's mind when one hears, for example, a "very high tone" without feeling that its location in musical intervals relative to another pitch is of interest. The voice pitch of a speaker is an example. An important difference between musical pitch and tone height is that all tones with a frequency within the audible range possess tone height, whereas musical pitch is a characteristic of tones falling within the frequency range of about 30 to 4,000 Hz.

Musically interested people have no reason to mistrust the notions of tone height and mel. They have been found to be quite efficient in explaining certain aspects of hearing, e.g., how vowel spectra are perceived. It has also been found that the mel scale is in much closer agreement with the frequency scale, which is represented on the basilar membrane in terms of the location of the maximum of the traveling wave. We will return to this in a later section.

III. Hearing

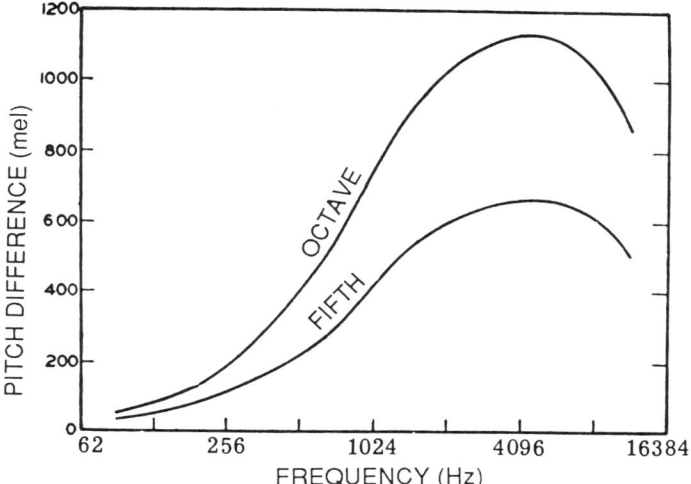

Figure 3.6. Sizes of an octave and a fifth interval, expressed as tone height difference, are shown as function of frequency. (After Stevens and Davis, 1938.)

4. Difference Limen for Tone Height

The just noticeable difference, or the *difference limen* (DL), is a term used for the smallest difference that can be perceived. The DL for frequency thus answers the question of how large a frequency change must be to be perceptible. It refers to the situation when two tones are presented in a sequence rather than simultaneously; in the latter case we know from Chapter 2 that beats are produced when tones of nearly identical frequency sound, and in such cases, it would be easier to listen for the beats than for a small difference in tone height.

Difference limen in frequency depends on several factors: frequency, intensity, and spectrum. However, the experimental conditions are also significant. For example, one can determine how wide a slow sinusoidal frequency variation needs to be to be noticeable. Figure 3.7 shows how wide such variations need to be at different frequencies and intensities when the frequency makes a few undulations per second. At 100 Hz, close to G2, the variations must be between 2.5 and 3.5%, or 2.5 and 3.5 Hz, to be perceptible. If the tone is soft, more are needed. At 1,000 Hz, the DL is between 0.5 and 1%, or 5 and 10 Hz. Thus, the discrimination power increases with rising frequency and with rising intensity. As a rule of thumb, for reasonably loud tones an audible frequency variation needs to be at least 3 Hz below 600 Hz, and for higher tones the amount is 0.5%.

DL is about 3% at approximately 100 Hz. This corresponds to an interval of a quartertone, i.e., half a semitone. It may appear strange that the smallest audible frequency difference should be that huge in the vicinity of the tone G2. But in practice it is not all that large under normal conditions. Up to now we have been considering sine tones, and for complex tones the DL is much smaller. One would imagine that if a complex tone contains partials above 600 Hz, a DL of 0.5% would be more realistic. This is slightly less than a tenth of a semitone.

For a complex tone it seems that the higher partials help the detection of small frequency differences. For the same reason it would be more difficult to perceive the exact pitch of a sine tone than of a complex tone with many strong partials. This may be one of the reasons why orchestras find it convenient to take the pitch reference from the oboe, which probably is the orchestra instrument with the strongest high harmonic overtones. A tuning fork does not have many overtones to brag about. The fact that it still is useful as a tuning reference depends on the fact that, in contrast to wind instruments, it is very insensitive to temperature changes. However, it serves its purpose as a frequency reference most efficiently if it sounds simultaneously with the tone to be tuned, so that beats reveal the tuning differences. For instance, if one presses a tuning fork against a piano and hits the A4 key, beats will tell if the piano has the same frequency as the fork.

It was mentioned above that the mel scale seems to reflect a property of the human hearing system. One support for this is the fact that the DL for slow frequency variations is almost constant for all frequencies if one expresses it in mel rather than in Hz, semitones, or percentage. It amounts to about 2.5 mel for reasonably loud tones.

DL for frequency is strongly dependent on the method of measurement, as mentioned. The results shown in Figure 3.7 refer to a sine tone, the frequency of which varied in time according to the pattern of a sine wave with the frequency

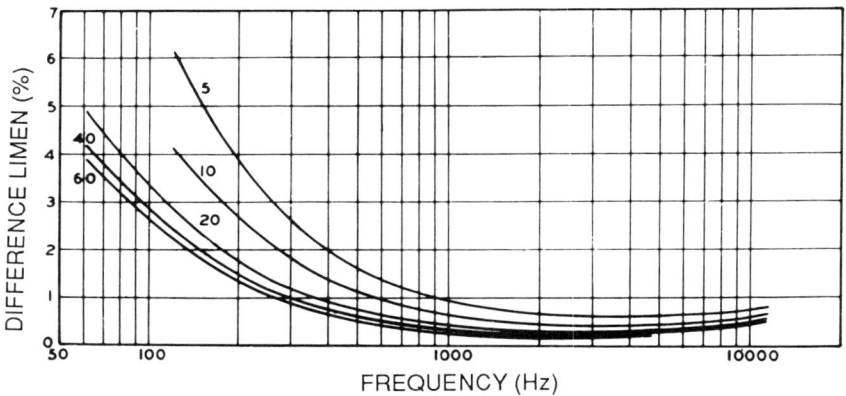

Figure 3.7 Smallest perceptible frequency variation at sound intensities shown according to measurements of Shower and Biddulph. (From Olson, 1968.)

of 4 Hz. Another method is to present tones in pairs that differ in frequency and ask subjects if the pitches are identical. Quite different DL values emerge from such experiments. In the bass they are no more than about a tenth of the corresponding values shown in Figure 3.7, but near 1,000 Hz they are close to what the figure shows. When the musicians in an orchestra adjust the tuning of their instruments in accordance with the oboe, it is probably this DL that applies.

5. Pitch and Vibrato

Several of the tones occurring in music are produced with a vibrato. This means that the tone is pulsating at a regular rate. This effect is achieved in two different ways. Either the fundamental frequency or the amplitude of the tone is undulating more or less sinusoidally around an average that remains rather constant. In some instruments, such as the human voice, the fundamental frequency variations affect the amplitudes of the partials, so that both frequency and amplitude vary.

Vibratos can differ in rate and depth, i.e., the number of undulations per second and the excursions from the average frequency or amplitude may vary. Provided that both rate and depth are within certain limits, a well-defined pitch is perceived. In the case of a frequency vibrato, the pitch perceived is practically identical with that perceived from a similar tone presented without vibrato and with a frequency corresponding to the average frequency. Thus, it appears that the ear computes the frequency average and uses this average as the basis for the pitch perceived. However, this process seems to work only if the rate is between 5 and 8 undulations per second, which is the range where the vibrato of the human singing voice is typically found. If the rate is slower, an undulating pitch is heard. It seems as if the hearing system is averaging the frequency information falling within a time window of a certain width; as long as there is room for approximately one complete undulation in this window, the average remains constant, but if there is space for only part of the undulation, a time-varying average results.

It is frequently assumed that the vibrato increases the DL for frequency: If the note has a vibrato, intonation is not all that critical. This seems true only to the extent that a vibrato hides beats. However, we seem to perceive the pitch of a vibrato tone with approximately the same accuracy as that of a vibrato-free tone.

6. Absolute Pitch

Some people develop a particular ability to recognize pitches so that they are able to tell what the pitch of a sounding tone is in almost any situation. This ability is called *absolute pitch*. There are reasons to believe that this capability is learned most easily in very early life.

Many regard this ability as an innate gift of God, vouchsafed only few. Others contend that absolute pitch is merely an example of a well-developed memory for pitches and that everyone can acquire an absolute pitch by sufficient training. Which of these two views is correct is hard to tell.

Absolute pitch is different from relative pitch, which definitely can be learned. The relative pitch implies the capability of identifying musical intervals between tones. Thus, everyone possessing a relative pitch is able to tell what interval a tone constitutes with a preceding tone. A good relative pitch is a must for all musicians.

There are a lot of different variants of absolute pitch. The most remarkable one and really the only true absolute pitch is that which allows a person to tell the name of any pitch heard and to produce correctly any pitch on dictation. There are stories that some singers possessed an absolute pitch so accurate that the orchestra used the singer rather than the oboe as its tuning reference. However, it appears somewhat doubtful that the absolute pitch can be accurate enough for such a purpose. Typically, the accuracy of the absolute pitch is in the vicinity of only 1%.

There are numerous substitutes for the true absolute pitch. Some use the lowest pitch they can sing, which is often reasonably stable, at least as long as the vocal folds are healthy; using this reference they can figure out the pitch name of a tone they hear. Others recognize how the throat "feels" when they sing a specific pitch; this kind of ability is frequently called *muscle memory* for pitch. Another method is to memorize the timbre of a certain pitch sung on a certain vowel. It also happens that musicians learn to recognize one single pitch that they repeatedly encounter. For example, some organists have heard the B flat used for giving the reference pitch to the minister in such copious quantities that they could recognize this single tone, and no other, wherever it appears and can produce it whenever asked. If one accepted also such cases of pitch recognition as examples of absolute pitch, it would be remarkable if not anyone could acquire one!

An absolute pitch is often an a nuisance for a musician, and it can certainly not be considered as a proof of an exceptionally high degree of musicality. Choir singers with an absolute pitch may have severe difficulties when singing in a key differing from that given in the score, whereas fellow singers do not even notice the difference. And keyboard instrumentalists with an absolute pitch may find it quite difficult to play on instruments tuned a semitone off as compared with the modern standard pitch. Further it sometimes happens that the absolute pitch changes when one gets old; for example, it may be quite disturbing to hear the famous Chopin waltz in F sharp major performed in F major.

B. Loudness and Amplitude

The loudness perceived of a sound is obviously closely related to its amplitude. It is this amplitude that directly determines the magnitude of the eardrum vibrations, and it would certainly have been very hard to understand if we heard sounds as deafeningly loud that were elicited by the most minute eardrum vibrations.

III. Hearing

The ability of the ear to process sound is nothing less than overwhelming. The amplitudes of the loudest sounds we can hear without perceiving pain is about ten million times greater than that of the softest sound we can perceive. Therefore, if we want to specify the loudness of sounds in terms of their amplitudes, we need a lot of numbers. But there is another, smarter way. We already saw that it is much simpler to specify *levels* using the logarithmic and relative decibel measure.

Before continuing and repeatedly making use of the notion of sound level in dB, it might be worthwhile to brush up our recollection of what, exactly, levels and dB stand for. The level of a sound is given in dB. The number of dB is obtained by taking the logarithm of the ratio between the sound pressure amplitude of the sound and that of a reference amplitude and then by multiplying by 20. One can also say that the number of Bels equals the log of the ratio between the intensity of the sound and a reference intensity. If the pressure of 0.00002 Pa is used for reference, the SPL is obtained. Table 3.1 lists sound pressure amplitudes and levels for different sounds.

1. Threshold of Hearing

Keeping the subtle construction of the ear in mind, it does not appear astonishing that the softest perceptible sound has different amplitudes at different frequencies. In other words, the ear's sensitivity varies with frequency. The SPL of the softest perceptible sound is called the *hearing threshold*. This threshold can be described by a curve varying with frequency, as shown in Figure 3.8. In the same figure the threshold of pain is also shown, i.e., the SPL of sounds that are so loud that they cause ear pain. Note that it varies between about 105 and 150 dB, depending on the frequency. The figure shows that the ear is most sensitive to frequencies in the range of 3–5 kHz. However, we are not very good at hearing sounds at low frequencies. To hear anything at the frequency of 32 Hz, cor-

Table 3.1 Approximate sound pressure amplitudes and SPL for different sounds

Sound source	Sound pressure amplitude (Pa)	SPL (dB)
Softest perceptible	0.00002	0
Soft rustling of leaves	0.00006	10
Whispering at 1 m	0.0002	20
Soft speech at 1 m	0.0006	30
Normal speech at 1 m	0.006	50
Lively street traffic	0.06	70
Pneumatic drill at 3 m	0.63	90
Discotheque	2.0	100
Threshold of pain	20	120

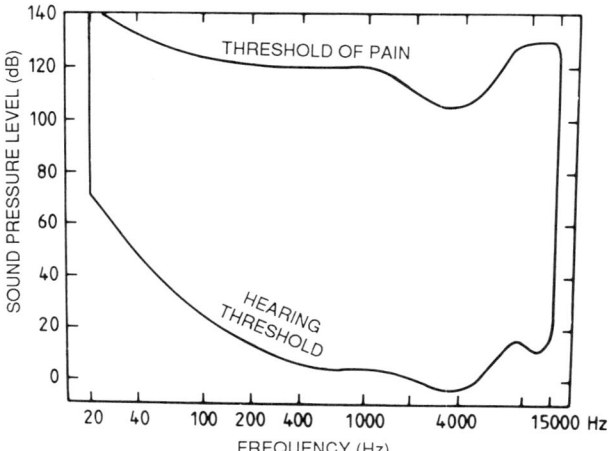

Figure 3.8 Audible frequency and sound level range.

responding to the pitch of C1, a substantial sound pressure is needed. In fact, the threshold at this frequency is midway between threshold of hearing and threshold of pain at 5 kHz!

Unfortunately, the threshold of hearing does not remain constant throughout life. It changes constantly, and above 20 years of age it rises slowly but continuously. This means that we cannot hear soft sounds as before. The hearing threshold also often rises if we hear very loud sounds for a long time. A warning sign of an imminent damage of the hearing system is that one senses a minor but temporary deafness after the loud sound stops.

A hearing damage implies that the hearing threshold is clearly higher than what is normal for one's age group. The damage can be specified in terms of the amount of decibels by which the threshold has increased above normal. Often such increases are limited to a certain frequency range.

A curve showing how the hearing threshold deviates from normal is called an *audiogram*. A hearing reduction of 40 dB at 3 kHz implies that an SPL of 40 dB above the normal is needed at this frequency to elicit a hearing sensation. Figure 3.9 shows an audiogram for a person with a hearing reduction that was probably acquired in his profession as an artillery man. The figure also illustrates a common method of determining the audiogram. One hears a soft sine tone with a slowly increasing frequency and decreasing sound level. When the subject can no longer hear the tone, he or she presses a button, and as long as the button is pressed the sound level increases. When the tone can be heard again, the subject is instructed to release the button, and then the sound level starts to decrease slowly again. At

III. Hearing 55

any moment the curve shows the sound level presented. This is the reason why the curve has its zig-zag form shown in the figure.

Risk of a permanent hearing reduction is affiliated with hearing very loud sounds, particularly sudden sounds as from explosions. To blow paper bags too close to the ear may not always be an innocent trick. The click sounds in the old telephone models produced very loud, sudden sound levels when the receiver was pressed against the ear. Harmonic musical sounds do not appear as detrimental as noise. In any event it has been determined that musicians in symphony orchestras are exposed to sound levels that are considered detrimental to the hearing, but still it appears that these musicians do not become more hard of hearing than people working in less loud professions. Among rock and pop musicians, however, the occurrence of hearing losses seems more frequent than normal. However, the results of these investigations of hearing losses among musicians are somewhat divergent.

In our time, music listening may not be completely harmless to the hearing system. It is easy to degenerate to the habit of using earphones at too high levels. The occurrence of hearing damages among young males inspected for military service in Norway was recently found to have doubled over the past 10 years. The earphone listening to music from portable tape recorders, "walkmans," is one possible cause for this. However, there are also other possible causes.

The shape of the hearing threshold changes continuously with age, as mentioned. Figure 3.10 illustrates this. We note that the hearing loss starts as early as

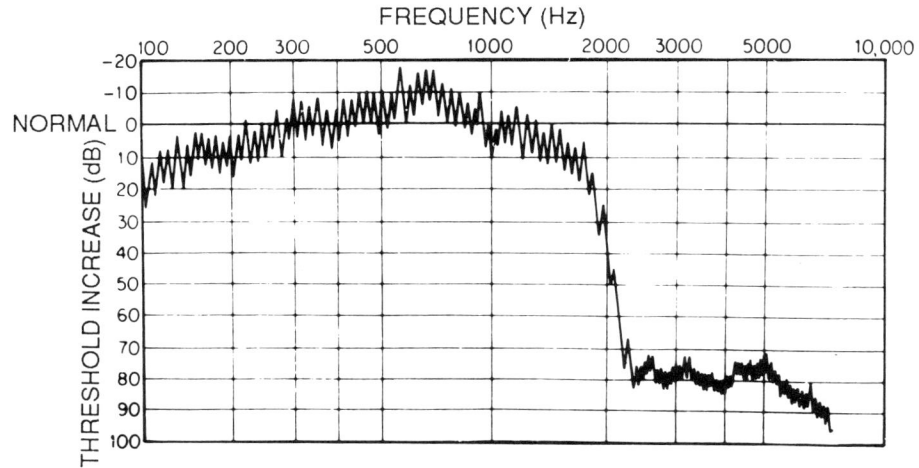

Figure 3.9 Audiogram for an artillery man with acquired hearing loss. (From von Békésy, 1960.)

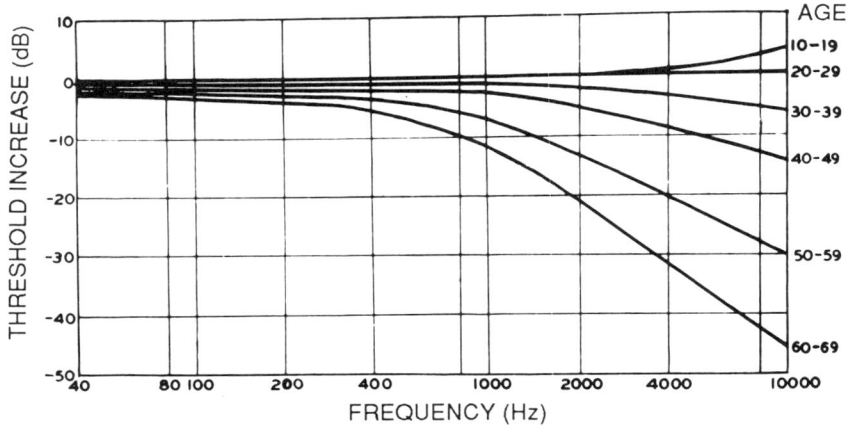

Figure 3.10 Dependence of the hearing threshold on age. (From Olson, 1968.)

the 20s. The hearing loss increases with increasing frequency. In the 50s the loss is about 20 dB at 4 kHz. The sound of crickets is in the vicinity of 8 kHz, so when we cannot hear these animals any longer we have acquired a considerable hearing loss in this frequency range. The highest sounds relevant to speech appears at about 7 kHz. Thus, difficulties in hearing speech need not occur even if we can no longer hear the crickets. What is difficult is to perceive speech in noisy surroundings. To succeed in this, a good directional hearing is needed, and this ability is strongly dependent on good hearing in the high-frequency range. Therefore, an early sign of a hearing loss at high frequencies is a frequent need to ask what was said in conversations in noisy surroundings such as restaurants and around happily chatting dinner tables. We will return to this issue at the end of this chapter.

It is not well understood how a hearing threshold shift affects music listening. Most relevant music sounds appear well below 10 kHz, but many instruments such as harpsichord and some percussion instruments contain significant components up to both 10 and 20 kHz. Elderly persons with a normal hearing loss for their age group therefore have difficulties in perceiving the highest timbral components in such instruments. It also appears that it would be difficult to trace what happens in a polyphonic piece of music, because the directional hearing would depend on this ability. However, it seems that the less we hear, the more efficiently we learn to use what we can still hear. For example, it can be mentioned that many extremely successful music teachers can continue to achieve very good results with students even when they have difficulties hearing both the phone and the doorbell. Thus, a very significant hearing loss does not need to exclude a person from the enjoyment of listening to music.

III. Hearing   57

2. Loudness Level and Phone

As the sensitivity of hearing is level-dependent, two tones of equal sound level mostly sound different in loudness. A sound level that is barely perceptible at 100 Hz produces a rather loud tone at 1,000 Hz. Therefore, sound level is far from being an ideal measure of perceived loudness.

A straightforward way of getting around this problem is to state that this tone sounds as loud as a sine tone of specified SPL at 1,000 Hz. Such a measure has been developed by experiments in which listeners adjusted the sound level of tones of different frequencies such that they matched the loudness perceived of 1,000-Hz tones with varied SPL. The measure resulting from these experiments has been called the *loudness level,* and the unit is called *phone*. We note that the loudness level in phone for a sound equals the SPL of an equally loud sine tone at 1,000 Hz.

Figure 3.11 presents equal-loudness contours. They show how the SPL of a tone must be changed with its frequency to keep its loudness constant. If a tone has a loudness level of 40 phones, it sounds as loud as a 1,000 Hz tone of 40-dB SPL. The strong frequency dependence is clearly illustrated.

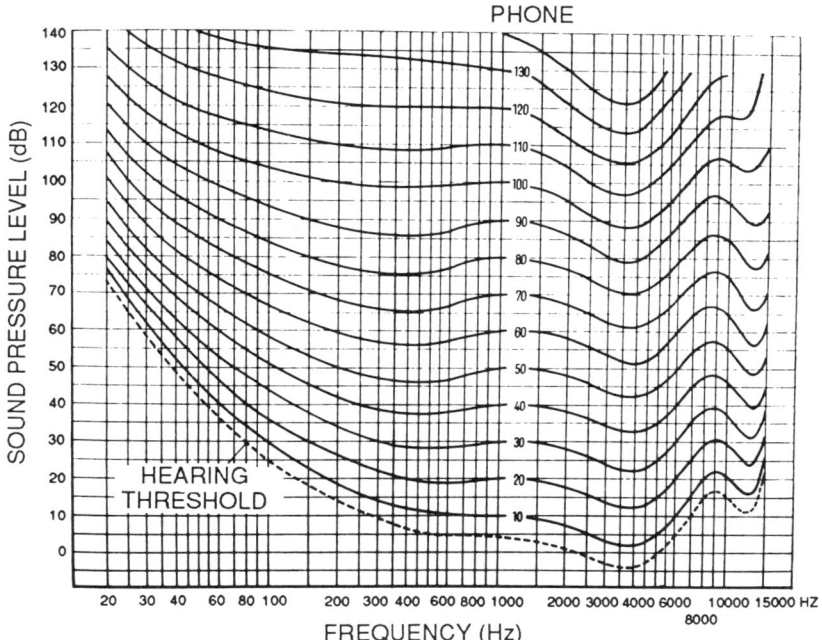

Figure 3.11 Equal-loudness contours showing how SPL of a sine tone must be changed with frequency to keep its loudness constant.

Again we observe that rather sizeable sound levels are needed in the bass and that the sound levels that are barely audible in the bass are quite loud at 1,000–4,000 Hz. We can make other observations, too. For soft tones, small increases of SPL mean a lot of loudness increase; the curves are densely packed near the hearing threshold. An increase from 0 to 70 phone at 500 Hz costs an SPL increase of 65 dB, but at 100 Hz the same loudness increase costs only 40 dB. If we regard reasonably loud tones above 100 Hz, the curves run rather parallel and do not bend upward much in the bass. When it comes to loud tones, 90 phone or louder, the curves are actually rather parallel. At high loudness levels above 100 Hz, the frequency dependence of the hearing system is not all that great.

The loudness level is a useful measure in the sense that equality in the loudness levels of two tones implies that these tones sound equally loud, regardless of their frequencies. The loudness level implies that the frequency dependence of the hearing system has been compensated for.

3. Loudness (Sone)

It would be a mistake, however, to believe that a 20-phone increase of the loudness level corresponds to the same increase of perceived loudness, regardless of whether this increase happens between 0 and 20 phone or between 80 and 100 phone. In fact, the significance of the absolute loudness is enormous. Therefore, there is a need for a different kind of loudness measure, and such a measure is offered by the *loudness* as measured in *sone*.

It was mentioned before (this chapter, Section III.A.3) that it is actually possible to quantify sensations. People produce reasonably similar experimental results when asked to tell what change corresponds to twice as high, long, loud, far, etc. This applies also to loudness. When subjects were asked to adjust the sound level of a tone such that it sounded as twice as loud or half as loud as another reference tone, reasonably similar data were obtained from different subjects. This was the method used when the loudness measure sone was developed. We observe the difference between (absolute) loudness and (relative) loudness level.

If two tones have the same loudness, this implies not only that they sound equally loud, but also have the same loudness level. It also implies that a doubling or halving of the loudness in sones will sound as if the sound became twice as loud or half as loud.

Figure 3.12 shows the relation between loudness and SPL. Different frequencies produce different curves, so frequency is the parameter in this figure. The reason for the very complicated picture is quite simple: the frequency dependence of the ear's sensitivity. If we take this into account, the simple curve shown in Figure 3.13 is obtained. It tells us that a 10-phone increase of the loudness level at soft sounds means a very small increase in loudness; an increase from 10 to 20 phone

III. Hearing

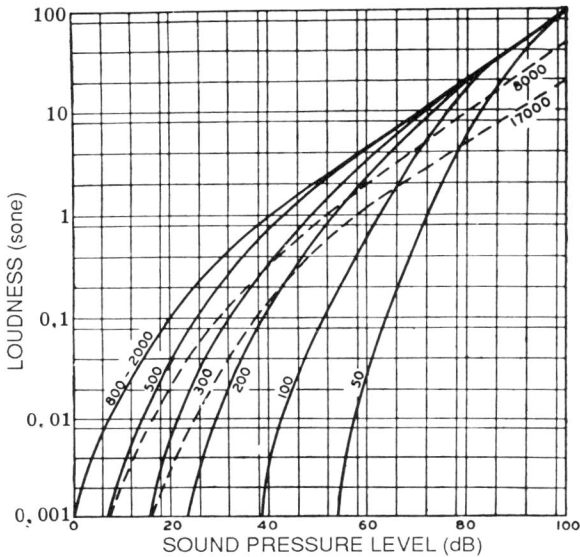

Figure 3.12 Relation between loudness as measured in sones and SPL, measured in dB, for given frequencies. (From Olson, 1968.)

yields no more than about 0.09 sone. However, an increase from 90 to 100 phone corresponds to no less than 40 sone. We can also note that for sounds louder than 40 phone, a 9-phone increase of loudness level doubles the loudness.

One might argue that sone must be the measure of loudness that is closest to perception. If so, one would expect the sone to be the perfect measure for the various dynamic signs that musicians deal with, such as mezzo forte and pianissimo. But this is far from being the case. It has been shown that wind instrument players tend to change the loudness by a constant number of decibels when they shift between two adjacent dynamic signs (Dekan 1972). This seems to demonstrate that, still, the decibel unit reflects a musically relevant aspect of perceived loudness. Loudness in sone certainly has a musical relevance, too, but not for distinguishing between dynamic signs.

4. Loudness Summation

It was mentioned that the sone unit is constructed in such a way that a doubling of the loudness in sones is perceived as a doubling of the loudness. Further it applies that if two tones, each with a loudness of 5 sones, sound simultaneously, the overall loudness of this compound is 10 sones. However, this principle of

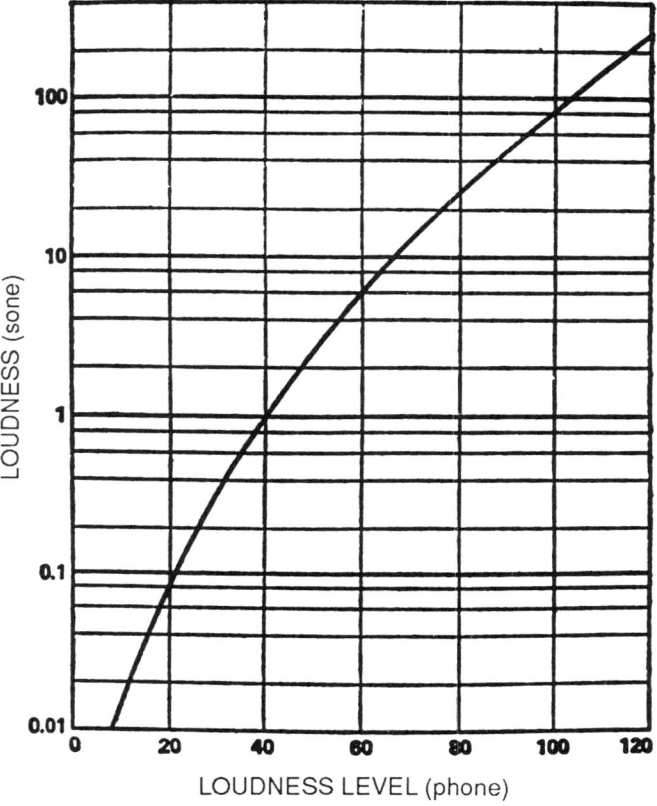

Figure 3.13 Relation between loudness level as measured in phones and loudness as measured in sones.

simple addition of loudnesses applies only under one specific condition, namely, that the tones are sufficiently separated in frequency. There is a critical frequency separation below which this simple addition of loudness does not apply. This separation is called the *critical bandwidth* of hearing. For bass notes it is about 100 Hz, and above approximately 400 Hz, it is close to a minor third. We will return in a later section of this chapter to this very important property of the hearing system.

If two tones with intensities I_1 and I_2 fall within the same critical bandwidth, the summed loudness is determined simply as the summed intensity I_t

$$I_t = I_1 + I_2 \tag{3.1}$$

III. Hearing

Thus, if the two tones have the same intensity, the total intensity is simply doubled in this case. If we regard the pressures, the truth is almost as simple. Intensity is proportional to pressure squared. If the sound pressure amplitudes of the tones are p_1 and p_2, the total sound pressure p_t

$$p_t^2 = p_1^2 + p_2^2 \tag{3.2a}$$

or

$$p_t = \sqrt{p_1^2 + p_2^2} \tag{3.2b}$$

If we put this very important equation into words, the truth sounds really enchanting:

The total sound pressure equals the square root of the sum of the partials' squared sound pressures.

This is a fascinating truth. Regardless of whether we consider pressure or intensity, we arrive at 3-dB increase by adding another tone of equal strength. We just noticed that it takes 9 phones to double the loudness, provided we consider reasonably loud tones only. Under the same conditions, 9 phones is not too far from 9 dB, as shown in Figure 3.11. This is much more than the 3 dB obtained if the tones are so close in frequency that they fall within the same critical band.

This is not only of academic interest. If one doubles the number of first violins in an orchestra, one theoretically obtains no more than 3 dB more in intensity, which is a very small increase in loudness. To double the loudness, an increase of 9 phones, which is close to 9 dB, is required. If 3 dB is obtained for each doubling, we need to multiply by 8 to arrive at 9 dB. Thus, 40 first violins should sound about twice as loud as five! However, if the violins are not playing in unison, but rather in different critical bands, a doubling of the number of violin parts produces a doubled loudness.

Organ players have probably noted this principle of increasing loudness for a long time. If one plays on two 8' stops rather than on one, the sound does not increase appreciably in loudness because the pipes in these stops sound with identical pitches. If, on the other hand, one adds a 4' stop to the 8' stop, there is a considerable loudness increase. One reason for this is that the 4' stop which is tuned one octave above the 8' stop adds some high-frequency partials that were not present in the sound of the 8' stop. There may be other reasons, too.

5. DL for Sound Level

How are small differences in sound level perceptible? That depends on the sound level. For reasonably loud sounds, the DL is near 0.5 dB, but closer to the threshold of hearing, it rises considerably, as Figure 3.14 shows.

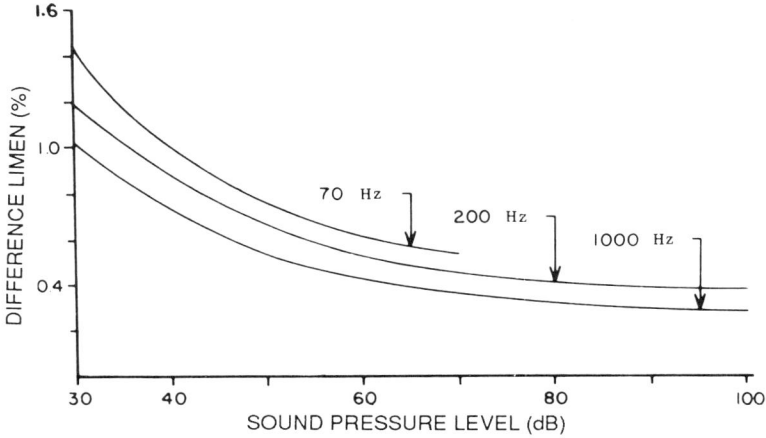

Figure 3.14 DL for sound level for sine tones plotted as function of SPL at three given frequencies. (From Backus, 1969.)

6. How Many Different Tones?

Our rather high sensitivity to level differences implies that there are wealthy possibilities to hear tones that differ in loudness. It can be shown that between the thresholds of hearing and pain there is space for about 280 tones of equal pitch but different in loudness. If we make the corresponding estimate for pitch, we arrive at the conclusion that there is space for about 1,400 tones between deepest bass and highest treble. How many tones can we distinguish between, then, taking both loudness and pitch differences into account? The answer is overwhelming: no less than about $280 \cdot 1,400 \approx 400,000$!* This is true for sine tones. If you find large numbers inspiring, you may also multiply this number by the number of timbres that we can distinguish. Then you arrive at numbers so large that you might believe that you are dealing with astronomy. One would be tempted to contend that these great numbers should arm composers with confidence that the possibilities of variation are practically unlimited in compositional work.

In practice, our hearing system is not all that generous. The tacit reasoning above is that we hear two tones in a sequence. Many of the melodic patterns in music depend on relations between tones appearing well-separated in time.

Another and perhaps more interesting point is how many degrees of pitch and loudness we can store in our memories, such that we can recognize and perhaps

---

*The sign $\approx$ means "approximately equal to". If you find it difficult to recall, it might help to think of it as an equality sign written with a hand trembling from doubt.

even name them. In this regard we are not at all as clever. In reality our memory seems to possess no more than approximately seven available slots! This seems to agree with the fact that there are seven pitches in the musical diatonic scale.* And musicians often work with six dynamic signs: pp, p, mp, mf, f, and ff.

C. Categorical Perception

Mostly we perceive differences in frequency, amplitude, and duration in the same way regardless of what the starting value of these magnitudes were. For example, a frequency change of 1% is similarly perceived, no matter if it appeared between 100 and 101 Hz or between 500 and 505 Hz. This kind of perception is called *continuous*.

In music and also in speech stimulus, changes are not always perceived in this way. Rather, great differences sometimes pass unnoticed and, under other circumstances, a very minor difference may sometimes make a very great difference to what we perceive.

Musical pitch intervals offer a typical example. Such intervals are measured in semitones or hundredths of semitones. Mostly, a small change of the size of an interval does not matter a great deal because musical intervals are sorted into classes or categories. When we hear musical intervals, we tend to sort them into such categories. Thus, an ascending fifth remains an ascending fifth even if it is tuned a bit narrow or wide. Maybe we note the deviation from the typical version by noting "That was a wide fifth!" In a listening test, one can present, for example, fifth intervals of varying widths and ask subjects to classify them. With growing width, subjects will continue to classify them as fifths for a long time. However, beyond a certain limit, the categorization flips so that suddenly most subjects start to perceive the interval as a minor sixth instead.

This kind of perception is called *categorical*. It is quite different from the continuous type, in which a certain stimulus change is always affecting perception in a similar way. Perception of pitch intervals is always categorical.

Perception of tone durations appearing in rhythmical sequences is another example. Musical listeners classify these with great skill into note values. The principles thereby applied are intuitive and far from being well understood. If one analyzes the physical durations, one finds that reality is confusing. For example, an eighth note may sometimes be played as long as a dotted eighth note without causing any doubt in the classification. The musical context is certainly a factor of great significance to the categorization. We will return to these issues in Chapter 10.

---

*Persons with an absolute pitch are clearly exceptional in this respect.

Figure 3.15 offers an example of how the boundary between the categories of fourth note plus eighth note and eighth note plus eighth note appeared in a laboratory experiment.

It is not only musical pitch and duration that are categorically perceived. One finds at least traces of the same ability in the perception of loudness, but formal listener experiments to demonstrate this have not yet been carried out.

Also the perception of the sound patterns that we recognize as individual speech sounds, or phonemes, is categorical. These sound patterns can be varied within wide limits without any change of the categorization, but a minor change beyond a magical limit has the effect that the sound is perceived as a different phoneme.

In general it appears that categorical perception is a very common phenomenon among people who are experts of judging certain aspects of things. Music listeners are experts in the interpretation of differences in musical pitch and duration. All of us who are acquainted with speech are experts in interpreting speech sounds. And also an expert of old furniture would certainly show something like a categorical perception of various characteristics of furniture styles. It seems that categorical perception is a particularly smart way of squeezing more information into one single perceptual dimension: Thanks to categorical perception, we may

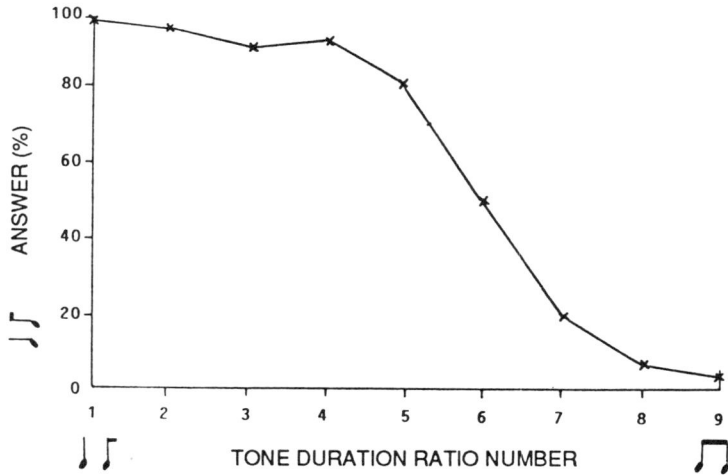

Figure 3.15 Categorization of note durations as function of tone durations in a tone pair. At the left extreme, the long note is 640 msec and the short is 320 msec, and at the right extreme both tones are 480 msec. Each step to the right on the scale corresponds to a decrease of the long tone and an increase of the short tone by 20 msec. Mostly, this small change of the durations does not cause much of a shift of categorization. However, the change from 540 + 420 to 520 + 440 msec (number 6 and 7 on the scale) produces a great effect on categorization. (From Clarke, 1987.)

III. Hearing   65

note not only that the new pitch is a fifth above the previous tone but also that it is a sharpened fifth. In a way, one dimension is carrying two messages simultaneously.

Another phenomenon is often affiliated with categorical perception, and indeed it is often regarded as a necessary condition for speaking about a case of "true" categorical perception: just at the boundary between two categories the discrimination ability is increased. This means that the DL is greater in the mid range of a category than at category boundaries. In the case of durations in a rhythmical sequence, this effect is weak. However, it does not seem overly obvious why this increase in discrimination power should be regarded as an inherent part of categorical perception.

D. Masking

In the past we have frequently been concerned with questions associated with the perception of simple sine tones. These, however, are rare in the real world, as mentioned. In reality it is only when we hear whistled tones or tones generated by electroacoustic devices that true or almost true sine tones are offered to our ears. How do complex tones affect our hearing then?

When two or more sine tones sound simultaneously as in complex tones, a phenomenon occurs that is of great significance for the timbre we perceive. This phenomenon is called *masking*. It implies that a tone becomes difficult to perceive because another tone is sounding. The tone is "drowned." Thus there are two agents associated with the phenomenon of masking, the *tone* and its *masker*. This masking effect can be measured in terms of the shift in hearing threshold for the tone: How many decibels louder than normal are needed to hear the tone in the presence of the masker?

The masking effect is generally greater above than below the frequency of the masker. A loud low-frequency masker efficiently masks a tone of somewhat higher frequency much more efficiently than a tone of slightly lower frequency. The louder the masker is, the longer its shadow reaches toward higher frequencies, as we might imagine. Figure 3.16a shows the masking effect in terms of the increase of the normal threshold of hearing for maskers of different loudness. The masker in this case was a narrow band noise centered at 415 Hz. The numbers on the curves show the sound level of the masker above its threshold value. The curves are called *masked thresholds*. They show the threshold of hearing in the presence of the masker, and they illustrate the fact that the masking is much greater above than below the masker frequency as long as the masker is loud. If, however, it is close to the threshold, the masking effect is more symmetrical.

These masked thresholds show a simpler course if the frequency scale is substituted by a scale that is more relevant to the properties of the hearing system.

Such a scale is provided by one in which each *critical band of hearing* is represented by the same distance. In addition, if instead of plotting the lowest perceptible SPL, one plots the shift in the hearing threshold, the graph becomes as simple and neat as is shown in Figure 3.16b. The figure shows that the softest perceptible level in the critical band of the masker is about 20 dB below the level of the masker. Above this band, the masked threshold falls off toward higher frequencies at a rate of about 5–13 dB per critical band, depending on the level of the masker. Toward lower frequencies, the roll-off is considerably steeper. In a graph of this type, the masked threshold can thus be approximated by means of three straight lines, provided that the masker is a sine tone that is neither too loud nor too soft.

When it comes to complex tones, it is no longer easy to keep track of which partial is a masker and which is a victim of masking. All partials mask each other and are masked by one another. The result is that partials in deep spectrum valleys are often completely masked by their stronger fellow partials. Figure 3.16c shows an example, being the vowel in *far*. Several partials above the peak at 1,000 Hz fall below the masked threshold and cannot be heard. Therefore, they do not contribute to the timbre of that sound. They can be eliminated from the spectrum without perceptible effect.

Most of the masking depends on the mechanism of the cochlea. Figure 3.3 showed that the vibration amplitude is greatest at a certain place on the basilar

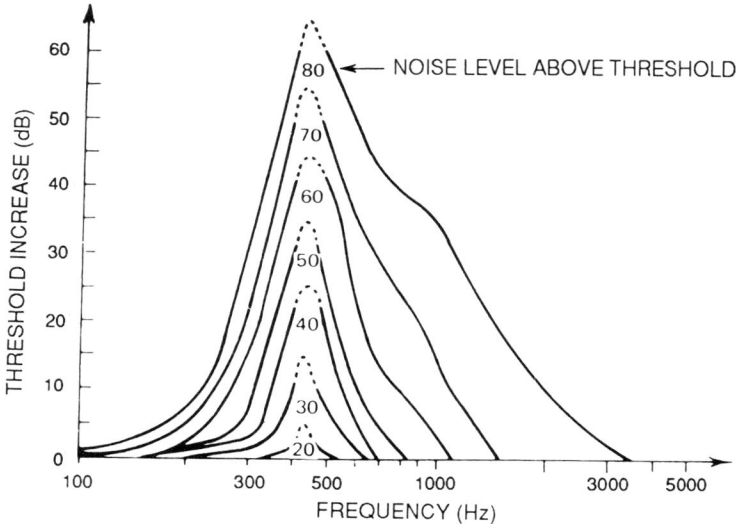

Figure 3.16a Masking effect of a narrow noise band masker centered at 415 Hz and presented at the given sound levels above the normal threshold measured in silence. (From Roederer, 1979.)

III. Hearing

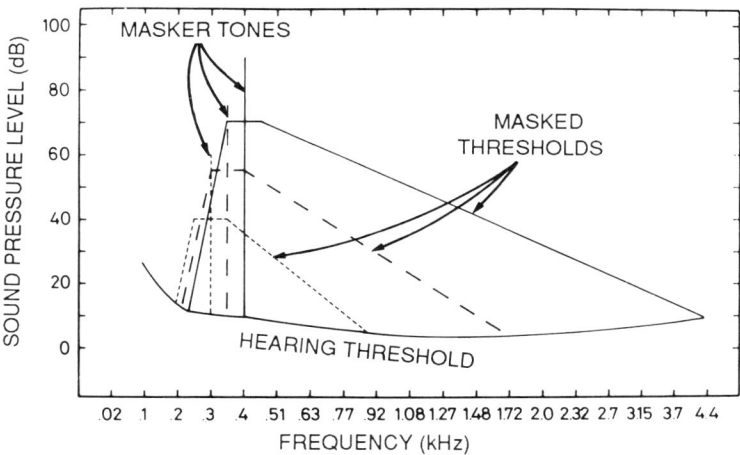

Figure 3.16b Idealized masking effects from masker sine tones at 300, 350, and 400 Hz presented at 50, 70, and 90 dB SPL. The frequency scale corresponds to the critical bands of hearing.

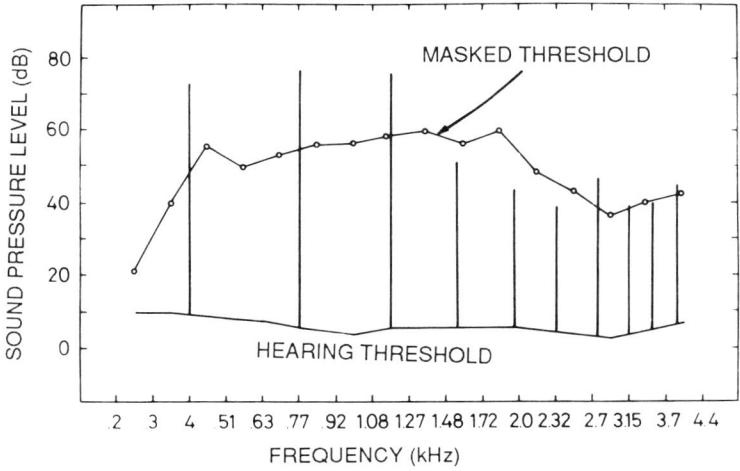

Figure 3.16c Masked threshold in presence of the spectrum shown, which corresponds to the vowel in the word *far*. The frequency scale is the same as in Figure 3.16b. Several partials fall below the masked threshold and thus do not contribute to the timbre.

membrane, when a sine tone is sounding, and the location of this place depends on the frequency of the tone. The figure also showed that the vibration amplitude decreases slower toward the oval window, i.e., toward higher frequencies, than toward the helicotrema, i.e., toward the low-frequency end. The vibration amplitude on the basilar membrane thus strikingly reminds us of the masked threshold curves. In fact, the masking phenomenon seems to depend strongly on these vibration amplitude patterns on the basilar membrane, and masking disappears nearly completely if the masker is fed to one ear and the tone to the opposite ear. This certainly belongs to the explanation of the fact that it is easier to separate a sound from a background noise in a stereo recording than in a mono recording.

How does masking agree with our experiences of listening to music? At first glance it does not seem to agree at all. On the contrary, it often seems easiest to perceive the highest part, the "melody part" in a polyphonic piece. The explanation to this phenomenon probably is that in almost all music instruments, the loudness increases with rising fundamental frequency. Therefore the top part often is the loudest. We must also remember the truth that Darwin launched, namely, that the fittest has the greatest chance of surviving. This principle certainly applies also to music. Instruments with properties preventing them from being heard as a melody part are not used for presenting melody parts by composers. After all, both composers and players possess a sense of hearing, and they intuitively know the properties of human hearing by heart. This intuitive knowledge on masking and other properties of hearing is significant to the way they shape the music.

In this connection one might spend some thoughts of playing music on authentic instruments. It is reasonable to preclude that the composer is the expert who can imagine very accurately how the instrument he is writing for is going to sound. The masking effects are decided by the sound levels of the instruments playing and their spectral characteristics. Constructing a new version of an existing instrument generally implies that its sound properties, and thus also its masking effects on other instruments, are changed. This may easily lead to difficulties in achieving audibility for effects that the composer counted on.

We have seen that the sound level has a decisive influence on the masking effect. This means that a polyphonic texture suffers from a loud acoustic realization. The weaker voices are less masked by the louder ones if one listens to music reproduced at realistic rather than exaggerated sound levels. And in, for example, a multi-voice organ composition, the point with the many voices is lost if the piece is played with organ stops that are too loud. Particularly if the bass part in the pedal is played too loudly, many of the higher voices will be masked.

For listening to many voices concerting with each other, it may be relevant that the ear perceives high notes quicker than low notes, and to be perceived it is important to be the first in the air or ear. The masking effect from a tone occurs almost exactly at the same time as the masker starts to sound, and shifts from one

III. Hearing 69

tone to the other often correspond to a very short amplitude reduction caused by tone onsets and/or micropauses, called *articulation*. It is enough that a soft tone is 20 or 30 msec ahead of its maskers for it to be clearly perceived, even in cases where it would be completely masked, if the tone and the maskers started simultaneously. It has been shown that in performed music accompanying voices often lag approximately by this amount of time, relative to the solo instruments. No wonder that ensemble music takes lots of practicing!

The ear also has a quite remarkable capability of filling missing parts of a sound signal. It may be enough that one hears short fragments of a tone some three or four times per second to have the illusion that the voice was sounding without interruption! Thus, by leading slightly in each tone onset, a soft instrument may be perceived as sounding continuously, provided that there are three or four onsets per second.

E. Combination Tones

When two sine tones excite the hearing system simultaneously, the ear produces new tones that we can sometimes discern along with the two primary tones. These tones, created by the ear, are called *combination tones*. There are different types of combination tones. Some occur always, whereas others occur only under certain conditions.

One type always occurs whenever two sine tones sound simultaneously. It has a frequency equal to two times the frequency of the lower tone minus one time that of the higher tone. There are also children of this type of combination tones; the common denominator is that the frequency of combination tone number n equals n plus one times the lower frequency minus n times the higher. If two tones with frequencies f_{high} and f_{low} sound, the nth combination tone occurs at

$$f_{c,n} = (n+1) \cdot f_{low} - n \cdot f_{high} \quad (n = 1,2,3,\ldots) \quad (3.3)$$

Let us assume that the frequencies f_{high} and f_{low} are 500 and 600 Hz. Then, combination tones number 1–4 have the frequencies

$$f_{k,1} = 2 \cdot 500 - 1 \cdot 600 = 1000 - 600 = 400$$
$$f_{k,2} = 3 \cdot 500 - 2 \cdot 600 = 1500 - 1200 = 300$$
$$f_{k,3} = 4 \cdot 500 - 3 \cdot 600 = 2000 - 1800 = 200$$
$$f_{k,4} = 5 \cdot 500 - 4 \cdot 600 = 2500 - 2400 = 100$$

We see that a harmonic spectrum is generated below the two primary tones at 500 and 600 Hz. The reason for this is that these two numbers have a common denominator, 100, which is the frequency separation between the combination tones in this spectrum. As long as one plays on instruments producing harmonic spectra, this kind of combination tone is rather harmless.

If the frequencies of the two primary tones do not have a common denominator reasonably close in frequency, the situation is different. Let us examine an example. Let us assume that f_{high} and f_{low} are 440 Hz and 500 Hz. We can see that

$$f_{k,1} = 2 \cdot 440 - 1 \cdot 500 = 880 - 500 = 380$$
$$f_{k,2} = 3 \cdot 440 - 2 \cdot 500 = 1320 - 1000 = 320$$
$$f_{k,3} = 4 \cdot 440 - 3 \cdot 500 = 1760 - 1500 = 260$$
$$f_{k,4} = 5 \cdot 440 - 4 \cdot 500 = 2200 - 2000 = 200$$

Again we see that combination tones appear at the multiples of the frequency difference between the primary tones, 60 Hz in this case. If the frequency difference is small, the combination tone frequency equals the beat frequency.

The combination tones appear below the frequency of the tones producing them. Therefore the primary tones cannot mask them so they are often easy to discern. A typical example is when two recorders play a narrow interval at high frequencies, for instance, the major third C5 – E5, or 523.3 and 660 Hz. The frequencies of the combination tones are $2 \cdot 523.3 - 660 = 386.6$ Hz, $3 \cdot 523.3 - 2 \cdot 660 = 249.9$ Hz, and $4 \cdot 523.3 - 3 \cdot 660 = 113.2$ Hz. These frequencies are close to but not equal to those corresponding to the pitches G4, B3, and a sharp A2 and they produce a rough timbre.

If one adds a bass note (e.g., C3), it will mask these ugly combination tones, so that they no longer disturb the harmony. Another method is to tune the major third perfectly just, $528 (= 4 \cdot 132)$ and $660 (= 5 \cdot 132)$. In that case the combination tones constitute a just triad $396 (= 2 \cdot 528 - 660)$, $264 (= 3 \cdot 528 - 2 \cdot 660)$, and $132 (= 4 \cdot 528 - 3 \cdot 660)$ Hz. This triad sounds thick but less rough than the previous one. In fact, it is probably these just triads between combination tones that render just intonation its charm.

F. Spectrum, Timbre, and Critical Bandwidth

There are reasonably straightforward relations both between pitch and fundamental frequency and between loudness and SPL. The relations between timbre and spectrum are much more involved.

Perhaps this is because timbre is really a wastebasket concept; in this basket everything is collected that separates the percepts of tones equal in pitch and loudness. This may go back to many different tone properties. The tones may differ with regard to the loudness envelope, the tone onset, decay, etc., but also, of course, with regard to spectrum characteristics such as the amplitudes of the partials. Using this definition of timbre, an onset difference must be counted as a timbre difference, and this obviously complicates the situation.

If two tones differ with respect to the amplitudes of the partials, it must be relevant to find out how this difference appears after it has been filtered by the auditory system. For example, the critical bandwidth mentioned above should be relevant. It is related to the distribution of vibration amplitudes along the basilar membrane. It has been found that it is perfectly feasible to account for timbre differences after the spectra have been filtered through the critical bands, i.e., after summation of all partials exciting the same critical band. For example, there will be no great timbre difference, if partial number 8 or 9 is strong in a spectrum, as these partials fall in the same critical band. And if one tries to quantify perceived timbre differences between complex tones, it turns out that these quantities are similar to those obtained when the sound level differences between the spectra in each of the different critical bands are summed.

As mentioned, the critical band is some kind of smallest frequency difference that will allow two tones to be perceptually identified as two autonomous tones rather than as one single buzzing unit. All partials falling within the same critical band form such buzzing percepts, which contribute to the timbre in a quite particular way that is easy to recognize. It has been given many different names, e.g., rough, grainy. A timbre that is void of this property is mostly described as soft, smooth, round, etc. Whether a timbre sounds rough or smooth depends mainly on how many partials excite the same critical bands and how loud these partials are. The louder two or more equally strong partials falling into the same critical band, and the greater the number of such partials, the greater the roughness of the timbre. All pairs of adjacent partials exciting the same critical bands contribute to the roughness of a complex tone, provided the partials are not too dissimilar in strength.

The critical bandwidth varies with its center frequency. For tones in the bass and up to about 500 Hz, it equals slightly less than 100 Hz. For higher frequencies, the critical bandwidth is about 20% of the center frequency. Thus, here it is not a constant number of Hertz but rather a constant frequency *ratio* approximately corresponding to a minor third. Measurements by various methods and the approximation mentioned of the critical bandwidth is shown in Figure 3.17.

Keeping the above in mind, it is interesting to find out how the partials of a harmonic spectrum fall along a critical band scale. If the fundamental is lower than 100 Hz, none of the partials has a private critical band. If, however, the fundamental is higher than 100 Hz, the situation is different. The fundamental will now reside in a band of its own, and the same is true for the lowest overtones, up to the fifth. But partial numbers 5 and 6 fall within the same critical band, and all higher pairs of adjacent partials as well, as the separation between them is narrower than a minor third. If a tone with a fundamental frequency above 100 Hz sounds rough, the reason is to be found among the partials above the fourth. If a

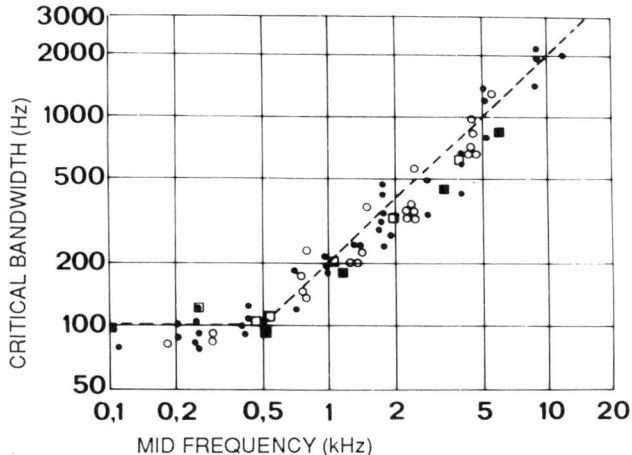

Figure 3.17 Width of critical bands of hearing as function of their center frequency. *Dashed line* shows an approximation: 100 Hz up to 500 Hz and a minor third for higher frequencies. *Symbols* refer to measurements obtained by various methods. (From Zwicker and Feldtkeller, 1967.)

tone below this fundamental frequency sounds rough, there is no wonder, because in this case all pairs of adjacent partials will contribute to the roughness. If it sounds smooth, some spectrum partials are likely to be missing, or some partials are much stronger than others. For example, the organ stop called Subbas 16' is frequently used for playing soft bass parts. Its timbre is very smooth despite the fact that its fundamental frequency goes as low as 32 Hz. The reason is that the even-numbered partials are missing in the spectrum.

Roughness is related to a phenomenon of fundamental significance to music theory, namely, *dissonance* and *consonance*. It should be observed that this consonance and dissonance are not identical to what music theorists have classified as consonance/dissonance, but rather the timbral quality *per se* of the sound of different musical intervals between simultaneously sounding tones. Systematic measurements have shown that the timbre of two simultaneous sine tones reaches maximum roughness or dissonance when the frequency difference amounts to a quarter of a critical bandwidth. Figure 3.18 shows how the perceived dissonance depends on the frequency separation as measured on a critical band scale. When the difference reaches one critical band or more, the dissonance quality is all but gone.

From the point of view of music theory, this appears to be nothing more than complete nonsense. A quarter of a critical bandwidth, which thus produces maximum dissonance, equals 25 Hz in the frequency range below approximately 500 Hz. This frequency separation corresponds to a major second if the lower tone is near 400 Hz, so that sounds fair enough. For higher tones, the situation is similar:

III. Hearing   73

There the critical band is about a minor third, and a quarter of that is in the vicinity of a minor second.

The disaster appears in the bass. A quarter of a critical band corresponds to a major third if the lower tone frequency is 100 Hz. Should a pure major third become dissonant simply because it is played in the bass? The same frequency separation of 25 Hz is a pure fifth if the lower tone is at 50 Hz, so then, a pure fifth is a dissonance! And what about major seventh; it should always be consonant except in the lowest bass, because it corresponds to a frequency difference wider than a critical band. However, systematic measurements show that people tend to find that an interval traditionally classified as "consonant" sounds progressively dissonant the farther down in the bass it is played. This is illustrated in Figure 3.19. In the very low bass, even the octave sounds dissonant!

Much of the mystery of these matters dissolves if one turns to complex harmonic tones rather than sine tones. Try to get an opportunity to listen to a just major third between two sine tones at 100 and 125 Hz. That interval pretty much growls. And then change the frequencies to the just minor seventh between 400 and 700 Hz. It sounds rather harmonious; the two tones combine without squabbling.

If, however, we turn to complex tones with harmonic spectra, the situation is entirely different. Then a great number of sine tones sound simultaneously, and

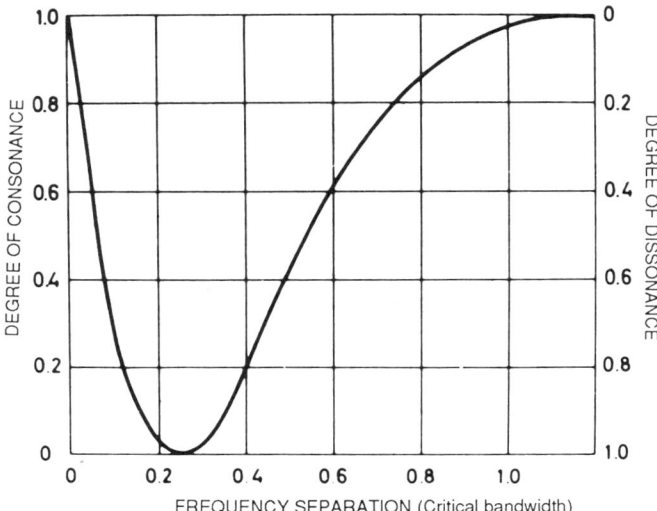

Figure 3.18 Relation between consonance/dissonance between two simultaneous sine tones and the critical bandwidth. When the frequency separation amounts to 0.25 critical band, the dissonance quality culminates, and when the separation exceeds one critical band, the timbre is completely consonant. (From Piomp and Levelt, 1965.)

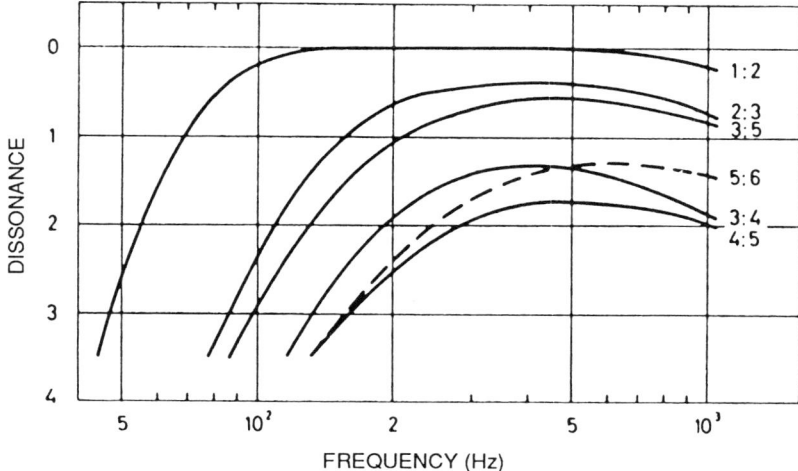

Figure 3.19 Degree of consonance of two simultaneous tones forming various musical intervals and played in various frequency ranges given in terms of the mid frequency. Dependence of the critical bandwidth has the effect that a consonant interval sounds more and more dissonant the farther down in the bass it is played. (From Plomp and Levelt, 1965.)

what determines the degree of dissonance is the sum of all the dissonances, which are produced by groups of sine tones falling into the same critical bands. It has been demonstrated that the dissonance theory for sine tones rather accurately predicts the dissonance perceived also between complex tones.

The degree of dissonance of two complex tones naturally increases if the most dissonant groups of sine tones are increased in amplitude. In a harpsichord, the lowest partials of the bass tones are not at all as loud as in a grand piano. This is the reason why a triad in compressed inversion in the bass sounds much more dissonant when played on a grand piano than on a harpsichord.

As mentioned, dissonance and consonance are traditionally regarded as two separate categories, according to which any interval can be classified. For example, the fifth has always been classified as a consonance, whereas a major third has been classified as dissonance during some periods and as a consonance during other periods. In reality, dissonance is one extreme of a dimension, the opposite extreme of which is consonance.

Any harmony is likely to slide along this dimension if the spectra involved are changed. If the lowest partials are made stronger, the interval will sound more dissonant. This leads us again, to the issue of performing music on authentic instruments. If we accept the certainly realistic assumption that composers are very musical people, the consequence will be that music should be played on authentic instruments. For instance, it would easily occur that the dissonances become too dissonant if the music is played on instruments with the low-frequency partials louder than what the composer had in mind. Some types of music survive the resulting sliding along the dissonance dimension very poorly. Imagine how piano music by Debussy would sound if played on a harpsichord!

In most instruments the timbre is far from being stable and constant. Depending on instrument and playing style, the timbre can be bent in various ways. The singing voice and the bowed instruments are extremely flexible in this sense. Further, several instruments can produce a vibrato, which mostly means that the fundamental can be varied up and down around a constant average. A spectrum of a single tone played on such instruments does not provide any exhaustive description of the timbre of that instrument. The spectrum of a sung vowel or a violin tone entirely depends on how the musician is playing the instrument, and the spectrum will depend on where it is taken in the vibrato cycle.

This is not true for all instruments, though. From the point of view of timbre, the organ is the least flexible instrument of all, so instruments with a very stable and constant timbre can still function as excellent music instruments. However, not even the tone from an organ pipe is completely stable. In the organ, as in many other blown instruments, the tone contains a small amount of noise. In small rooms or at a short distance from the instrument, it is often easy to hear this noise, at least if one listens for it.

It is sometimes maintained that the timbre that one perceives from a sound is substantially influenced by the tone onset. This is true only in the sense that the onset is very important when we try to *recognize* instruments from the sound. The reason for this is quite simple: The onset is mostly very characteristic. The effect is often quite striking. If one deletes the original onset of a tone from an oboe, it is almost impossible to hear what instrument is playing, and if the first few milliseconds are deleted in a piano tone, it sounds more like a harp!

The great relevance that our perceptive system attaches to the onset is hardly curious. A spectrum of the sustained note is much less reliable as a characteristic, because it is highly influenced by the standing waves of the room. The onset, on the other hand, arrives undistorted to the ears.

The decay is also perceptually important. Plucked instruments such as the lute and the guitar are often recognized from the decay.

However, the timbre *per se* is not affected by the onset to any great extent. As long as the spectrum is the same, the timbre sounds the same, no matter what tone onset preceded it. But the onset affects the sound image as a whole in the sense that it is significant for what instrument we believe is playing.

G. Hearing Directions

How can we hear that a sound comes from a certain direction? This ability, which is called *directional hearing,* has a quite decisive influence on our ability to separate the important from the irrelevant in the chaos of sound signals that strike our ears. This ability builds on a number of different effects which our perception intelligently takes advantage of.

One important aspect of directional hearing is the distance to the sound source. At first glance one might think that it is the sound level that gives us a hint. However, to be able to use this clue we need to know the original sound; otherwise it is hard to tell a loud distant sound from a soft close sound: Both sound soft. In the spectra of the human voice as well as most other music instruments, the higher overtones gain more sound level than the lower ones when loudness is raised so the spectrum balance may work as a clue. Still, it turns out that it is a little more difficult to tell the distance for unknown voices than for voices of friends!

A more reliable alternative is our capability to use the difference in relative sound level between the direct sound and its first reflexes. If the direct sound is much louder, the distance sounds short. This effect is dependent on the size of the room, but it is no problem to get an idea about the room size, e.g., by simply taking a look.

Distance is one dimension, and direction is another. The perception of the location in the left-to-right dimension makes use of several different effects. One is actually the tone onset. The distance between the ears as measured around the skull is about 20 cm. A sound originating on the axis of the right ear canal has to travel a 20-cm-longer distance before it reaches the left ear. This delays the arrival time by about 0.6 msec. Our perceptual system makes use of this tiny little arrival time difference for telling us where in the left-to-right dimension the sound originated. This strategy works only for sounds with short, distinct onsets, as we might expect.

Another effect helping us to place the sound source in the left-to-right dimension is the phase difference between the two ears' versions of the same sound. The phase difference arises for the same reason as the arrival time difference just discussed or the travel time difference between the ears. It works only for frequencies below 1,500 Hz, because the neural system cannot keep track of phase information at higher frequencies.

For higher frequencies, the ear uses loudness differences between the ears for the localization in the left-to-right dimension. This criterion is useful both for short click sounds and for long tones, but it is important only for frequencies above 1 or 2 kHz. The reason for this is good: It is only high frequencies that cannot swim around the skull and that therefore get lost or attenuated on their way, thus producing a weaker sound in the far ear.

For locating the sound in the front–back dimension, a quite different effect is used, namely, one that occurs when the sound is washing through all the little remarkable ridges and valleys in the outer ear. This effect appears at high frequencies, 5 kHz and above.

Summarizing, we see that the high frequencies are quite important in sound localization. When the aging of the hearing system has raised the hearing threshold for these high frequencies, the capability of sound localization is suffering. The most noticeable effect is that it becomes difficult to tell signals from competing background noises. Cocktail parties are not as much fun as before, because you can no longer hear what people are saying.

When locating sounds under normal conditions, we also take some help by moving the head, and in this way we can assemble some more criteria for the localization. Further, most of us are happily fitted with both eyes and ears, and whenever it is easier to get clearer information from the eyes, we do not hesitate to make use of them. A combination of ear and eye is often very efficient, and in such cases that is used. Humans are indeed smart; we take whatever is at hand that is useful without asking too many questions regarding if the criteria were chosen systematically or logically. Actually the choice is quite whimsical in many instances but it is nearly always the most reliable alternative that is preferred.

A particular aspect of directional hearing is the fact that sounds always appear as if they came from the direction from which the first sound arrived; later reflections do not change the perceived direction. This effect is referred to as the *precedence effect*. It is highly relevant when loudspeakers are placed in large rooms. As the sound travels faster when converted to electrical signals, the loudspeaker sound reaches the listener earlier than the airborne sound if no special arrangements are made. Then the precedence effect makes the sounds sound as if they came from the loudspeaker. To avoid this, the electrical pathway is mostly provided with some delay circuits, which are adjusted so that the airborne sound reaches the listener first. Even in cases in which the airborne sound is much weaker than the loudspeaker sound, the effect remains so that the sound image is coming from the real sound source rather than from the loudspeaker.

This completes our presentation of the ear and the hearing system. It took us some time and a good many pages, but there were quite a few things of musical interest to talk about. For those who feel like getting more profound information about these matters, there are many books to read. Further, as mentioned before, we will return to some musically relevant and more sophisticated capabilities of the hearing system in Chapter 10.

CHAPTER 4

Scales, Tunings, and Temperaments

I. INTRODUCTION

For reasons explained in the last chapter, it is necessary to distinguish between different types of pitch. One type, the musical pitch, is that kind of pitch that allows us to place tones in a musical scale. The other type of pitch, the extra-musical pitch or tone height, is the one that you think of when listening to the pitch of a speaking voice, i.e., that kind of pitch that is hard to place into a musical scale.

There is a strange thing with musical pitch. In a sense, it does not increase continually when frequency rises. Two tones one ctave apart share some aspect of musical pitch; with regard to this aspect they are identical. If one plays an ascending diatonic scale, one gets the idea that the musical pitch of the eighth note in the scale is the same as that of the starting note. This aspect of musical pitch is called *chroma*. The chroma of scale tones are identical for tones one octave apart. The chroma of a note can be specified in terms of scale tone names such as sol or D♭.

Because of the cyclical nature of chroma, it is enough to construct a scale of musical chromas encompassing no more than one octave. If that scale is simply replicated one octave higher or lower, the same series of chromas are obtained in the new octaves. In this chapter we will see how different series of chromas, i.e., musical scales, can be constructed.

II. SOME WORDS ON THE ART OF COMPUTING

As we will see later, any musical interval such as an octave or fifth corresponds to one and only one frequency ratio. For example, an octave corresponds to the ratio of 2:1, so that a doubling of the frequency raises the pitch by one octave. Similarly, other intervals correspond to other frequency ratios.

Ratios are therefore quite essential in the following, so the reader will have many reasons to use her or his abilities in calculation, particularly to handle

II. Some Words on the Art of Computing

mathematical ratios. Next, we will spend a few moments recapitulating how this is done, encouraging readers who are confident with this handicraft to skip this section.

A ratio is a mathematical expression that can be written a/b, where a and b are numbers. The number to the left of the slash is called the *numerator*, and the number to the right is called the *denominator*. The ratio a/b can be written in different ways:

$$\frac{a}{b} = a/b = a:b \tag{4.1}$$

It is always possible to expand a ratio, and sometimes it can be reduced. This implies that one multiplies or divides both the numerator and the denominator by the same number:

$$\frac{4}{5} = \frac{2 \cdot 4}{2 \cdot 5} = \frac{8}{10} \quad \text{(expanding)} \tag{4.2}$$

$$\frac{4}{6} = \frac{2 \cdot 2}{2 \cdot 3} = \frac{2}{3} \quad \text{(reducing)} \tag{4.3}$$

"The number a relates to number b as 4 to 5." This sentence is actually nothing less than an equation:

$$\frac{a}{b} = \frac{4}{5} \tag{4.4}$$

Note that "relates to" is written as a slash sign and that "as" corresponds to an equality sign. This translation is the key to all calculations of frequencies for tones that constitute certain intervals.

Equation (4.4) can be written in many different ways:

$$\frac{a}{b} = \frac{4}{5} \tag{4.4a}$$

$$\frac{5a}{b} = 4 \tag{4.4b}$$

$$\frac{a}{4b} = \frac{1}{5} \tag{4.4c}$$

$$a = \frac{4b}{5} \tag{4.4d}$$

$$b = \frac{5a}{4} \tag{4.4e}$$

$$\frac{5}{4} = \frac{b}{a} \tag{4.4f}$$

$$\frac{5a}{4b} = 1 \tag{4.4g}$$

$$\frac{5}{4b} = \frac{1}{a} \tag{4.4h}$$

$$\frac{4}{5a} = \frac{1}{b} \tag{4.4i}$$

The numbers that a and b represent in the expressions above may be integers as well as fractions. Note that if we want to multiply a/b by c/d, the following applies:

$$\frac{a}{b} \cdot \frac{c}{d} = \frac{a \cdot c}{b \cdot d} \qquad (4.5)$$

Say that

$$a = 2 \quad b = 3 \quad c = 5 \quad d = 6$$

Then

$$\frac{2}{3} \cdot \frac{5}{6} = \frac{2 \cdot 5}{3 \cdot 6} = \frac{10}{18} \qquad (4.6a)$$

which can be reduced:

$$\frac{10}{18} = \frac{5}{9} \qquad (4.6b)$$

If the ratio a/b should be divided by c/d, we get

$$\frac{a}{b} \bigg/ \frac{c}{d} = \frac{a \cdot d}{b \cdot c} \qquad (4.7)$$

Say that a, b, c, and d are still equal to 2, 3, 4, and 5, respectively. Then,

$$\frac{2}{3} \bigg/ \frac{5}{6} = \frac{2 \cdot 6}{3 \cdot 5} = \frac{12}{15} = \frac{4}{5} \qquad (4.8)$$

It is quite necessary to master all these simple rules and routines to keep the rest of this chapter from being an overly boring and time-consuming reading.

III. WHAT IS AN INTERVAL?

We mentioned before that intervals correspond to frequency ratios, so that, for example, tones having the frequency ratio of 1:2 sound as an octave. Other intervals obey the same principle. We repeat this immensely memorable truth:

Intervals Correspond to Frequency Ratios

This agrees nicely with the fact that an interval can be transposed in octaves. Two tones with the frequencies of 200 and 300 Hz produce a fifth. This means that the frequency ratio for the fifth is

$$\frac{300}{200} = \frac{3}{2} \qquad (4.9)$$

IV. Calculating Intervals

Let us now transpose the tones one octave downward. Then their frequencies must be halved, according to the above. We get the frequencies 100 and 150 Hz. The ratio between these tones is

$$\frac{150}{100} = \frac{15}{10} = \frac{3}{2} \tag{4.10}$$

i.e., the frequency ratio remained the same.

We must observe that an interval corresponds to a *ratio* between the fundamental frequencies, whereas the *differences* between these frequencies lack relevance for the interval. This is hardly remarkable. The difference obviously depends on what octave the interval is played in; it was 100 and 50 Hz in our example. But the ratio, as the interval, does not depend on the octave; it was 2/3 in both cases. This should suffice as a reminder of the very important principle that a musical interval between two tones corresponds to the ratio between the tones' fundamental frequencies.

We mentioned earlier that musical instruments produce complex tones; thus they contain an entire series of partials. The pitch perceived from a complex tone corresponds to its fundamental frequency. The interval between two complex tones corresponds to their fundamental frequency ratio. For the sake of brevity, we will henceforth use the term *frequency ratio* instead of *fundamental frequency ratio*.

IV. CALCULATING INTERVALS

As an interval corresponds to a frequency ratio, one can calculate the frequency of a tone that will produce a specific interval with a reference tone of known frequency, provided the frequency ratio of the interval is also known. Assume that we want to compute the frequency of a tone one fifth above the pitch of A4 with the frequency of 440 Hz. The frequency ratio of the fifth is 3:2, as we will see soon. Let us call the frequency of the pitch of E5 X Hz. Then, 440 relates to X as 2 relates to 3, i.e.,

$$\frac{440}{X} = \frac{2}{3} \tag{4.11}$$

$$X = \frac{3 \cdot 440}{2} = 440 \cdot \frac{3}{2} = 660 \text{ Hz}$$

Note that the frequency X is higher than 440; therefore, both X and 3 are below the line. The pitch of A4 has a lower fifth also. Hence, its frequency must be lower than 440 Hz. Let us assume that the frequency of the pitch of D4 has the frequency of Y Hz. Then, 440 relates to Y as 3 relates to 2, or

$$\frac{440}{Y} = \frac{3}{2} \qquad (4.12)$$

$$Y = \frac{2 \cdot 440}{3} = 440 \cdot \frac{2}{3} = \frac{880}{3} = 293.33\ldots \text{ Hz}$$

Note that we had to let the numerator and denominator switch places in the frequency ratio for the fith to arrive at the correct frequency. This is reasonable. If the figure in the numerator is greater than that of the denominator to the left of the equality sign in Eq. (4.12), the same must apply also to the expression to the right of this sign.

There is another thing, too, that is worthwhile noticing. We saw that to calculate the frequency of a tone at a cetain interval's distance from another tone with known frequency, we had to multiply that frequency with the frequency ratio of the interval. In Eqs. (4.11) and (4.12) we multiplied 400 by the frequency ratio of an ascending and descending fifth, or 3:2 and 2:3, respectively.

This is fundamental so let us summarize:

To calculate the frequency of a tone at a certain interval's distance from another tone with known frequency, multiply the frequency of the known tone with the frequency ratio of the interval. The higher number must be on the same side of the line on both sides of the equality sign.

It is also worthwhile to point out that every frequency ratio written in the form $a{:}b$ can also be written $c{:}1$. If we prefer this form, we simply have to divide a by b. For example, 3:2 can be written as 1.5:1, and 2:3 as 0.6666...:1. These ratios are equivalent to the integer ratios mentioned before, although the accuracy is sometimes dependent on the number of decimals. Many times it might be simpler to multiply by, for example, 1.5 than with 3:2.

As mentioned, every interval corresponds to a frequency ratio. How can we then remember what ratio corresponds to what interval? The simplest is to memorize the intervals between the frequencies of the partials in a harmonic spectrum. In Chapter 2 we saw that these frequencies correspond to a harmonic series. If the frequency of the fundamental is f_1, the partial number n has the frequency $n \cdot f_1$. The harmonic series for the pitch of A1 written on the classical music staff is shown in Figure 4.1.

The frequencies of the partials as well as the intervals between them and the frequency ratios are illustrated below the staff. The middle line shows the frequency ratios between neighbors in the series. They look very simple indeed. All of them are of the type $(n+1)/n$, where n is an integer. The reason for this is simply the fact that these frequencies constitute a harmonic series. Let us contemplate an example:

$$\frac{275}{220} = \frac{5 \cdot 55}{4 \cdot 55} = \frac{5}{4}$$

if we reduce. The intervals that occur between neighbors in a series of harmonic partials are called *just, pure,* or *harmonic*.

V. Consonance and Dissonance

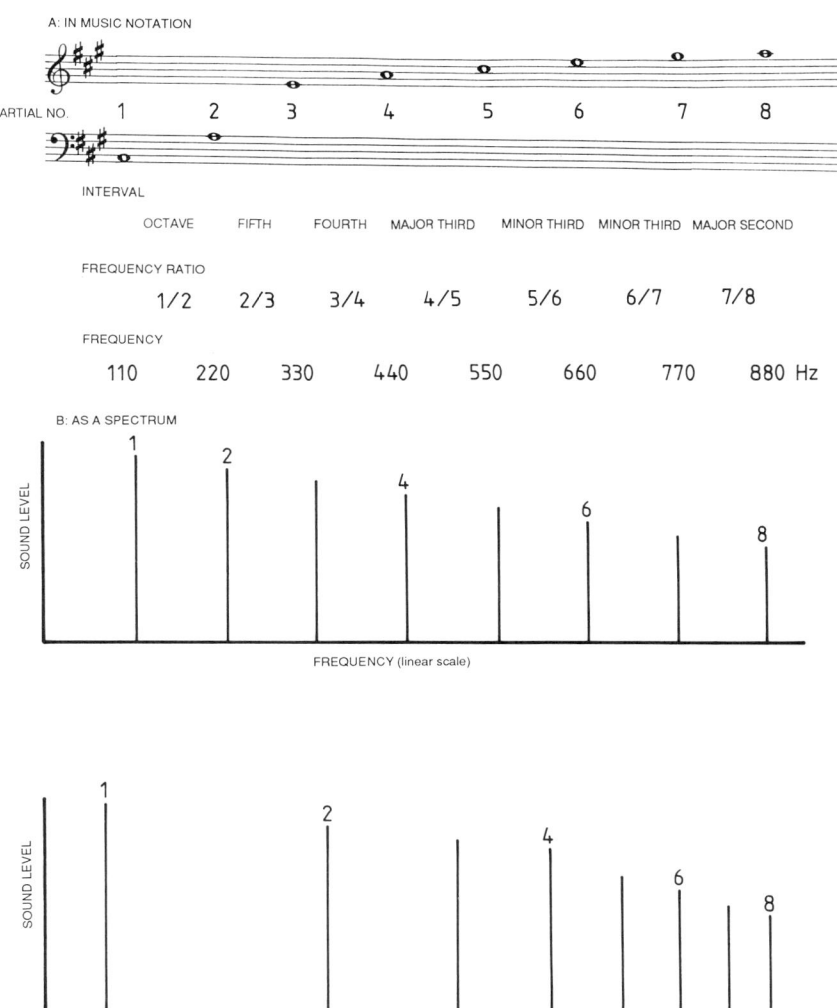

Figure 4.1 Harmonic overtones series.

V. CONSONANCE AND DISSONANCE

Before turning to the demonstration of how the frequencies of scale tones can be calculated, it may be pertinent to take a bit closer look than was made in Chapter 3 on the fundamental perceptual qualities called *dissonance* and *consonance*.

If two sine tones sound simultaneously, the frequencies of which differ by some Hz, beats occur. If the frequency difference is somewhat higher than 10 Hz, the sound receives a special quality; the tones melt together into a buzzing, rough, or dissonant timbre. In fact, dissonant really means "sounding apart." If the frequency difference is considerably greater than 10 Hz, the interval sounds more consonant, which really means "co-sounding." It is as if both tones then had regained their autonomy; they do not fight each other any longer.

Before continuing it is appropriate to present an important distinction between two different kinds of intervals. A *dyad* is an interval between two simultaneously sounding tones. Thus, the sound quality of a dyad may contain beats and roughness. This is not the case with intervals between successively sounding tones. Such intervals will be referred to as *melodic*. However, when it does not matter how the tones appear, we will continue to use the term *interval*.

In the previous chapter we saw that the critical bandwidth has a very important influence on what sounds dissonant and consonant. The critical bandwidth reflects a characteristic of the construction of the inner ear. For tones in the bass region and up to about 500 Hz, the critical bandwidth is slightly less than 100 Hz, and for higher frequencies it is close to a minor third. Two equally loud sine tones never sound dissonant if their frequency separation is greater than a critical bandwidth. If the frequency separation equals a quarter of a critical bandwidth, the dissonance is maximum.

When two complex tones with harmonic spectra sound simultaneously, a number of sine tones sound. The dissonance of this ensemble of tones depends on the magnitudes of the frequency distances between neighbor partials as measured on a critical band scale. The degree of dissonance is determined by distance, as measured in critical bands, between these partials and their strengths. The decisive factor is how many strong partials fall within the same critical bands. The more such partials, the higher the degree of dissonance.

If the number of dissonating partials is reduced, the result is a less dissonant quality. When two complex tones are sounding, the number of dissonating partials can still be low if the tones have many partials in common. It turns out that harmonic tones with fundamental frequencies that can be expressed as ratios between small integers share many partials. These are the intervals called *just*. The lower the integers, the higher the number of common partials. Beats occur in dyads, when the tones constitute an almost but not a perfectly just interval, because in this case partials that differ by few Hertz in frequency sound simultaneously.

The degree of consonance of dyads between two complex tones having harmonic spectra thus mainly depends upon the frequency ratio. An octave is the most consonant dyad possible: Here, all partials of the upper note are common with partials of the lower note. Let us confirm this by figuring out what the frequencies of the partials will be in case the fundamental frequencies are at 100 and 200 Hz:

Lower note: 100, 200, 300, 400, 500, 600, 700, 800, ... Hz
Higher note: 200,   400,   600,   800, ... Hz

V. Consonance and Dissonance

The second most consonant dyad is the fifth with the frequency ratio of 2:3.

Lower note: 200, 400, 600, 800, 1000, 1200, ... Hz
Higher note: 300,   600,   900,   1200, ... Hz

We conclude that if the interval has the frequency ratio of 1:2, every *second* partial of the lower note is common with *every* partial of the higher note. If the frequency ratio is 2:3, every *third* partial of the lower note agrees with every *second* partial of the higher note. This is the relation between consonance and frequency ratios for harmonic spectra.

An interesting consequence of the significance of the critical bandwidth is that the degree of dissonance of a dyad depends on many factors other than the frequency ratio. If the tones have many strong overtones, the consonance quality is reduced. A consonant dyad becomes increasingly dissonant when it is transposed downward on the frequency scale, just as for sine tones. For example, a major third sounds reasonably consonant around A4, but if played close to C2 it sounds quite dissonant on most instruments. The relation between consonance and frequency ratios is also entirely dependent on whether the tones have harmonic spectra. Consonance is apparently a highly conditioned phenomenon. It is stimulating to realize that the dissonance/consonance concept in music theory would have been entirely different if our musical instruments had not provided us with harmonic spectra!

It should also be observed that consonance is a quality that can be transformed *gradually* into its opposite, dissonance. Therefore one dyad may be more consonant than another one. This is not to say that it is impossible to maintain a classification of dyads in terms of consonance and dissonance, as has been done in music theory. With time, however, the boundaries between these classes have moved around a bit. The major third has been member of both the dissonance class and the consonance class in different periods.

When we henceforth speak about just intervals or dyads, we mean dyads that do not produce beats when played on instruments with harmonic spectra. They normally, but not invariably, sound consonant. Their frequency ratios can be expressed in terms of small integers.

What is the relation between consonance/dissonance and scales? In traditional tonal music, changes of the degree of consonance or dissonance have obviously been used as means for musical expression. Composers have varied the degree of consonance in a meaningful way. This made it desireable to use a pitch system or scale, which contained both very dissonant and very consonant intervals between the tones. This desire has been realized in basically one single way in our traditional Western concert music culture: the diatonic scale. We will explain this in more detail later in this chapter.

There are different ways in which the tones in this scale can be tuned. We will call these solutions *tunings*. Three such tunings will be explained in the next sections. The common feature of the scales to be presented is that they all contain the octave, the most consonant interval of all if we restrict ourselves to tones with harmonic spectra. The task is therefore to find ways of dividing the octave that offer different degrees of consonance/dissonance, ranging from very consonant to very dissonant.

VI. PYTHAGOREAN TUNING

The so-called Pythagorean tuning divides the octave by applying a very simple principle: One computes the frequencies of a number of tones along the circle of fifths simply by repeatedly using the frequency ratio of the fifth, or 2:3. Then the tones are transposed in octaves, such that they all arrive in the same octave. It is, of course, also possible to reach the same result by alternately using the fifth and its inversion, the fourth with the frequency ratio of 3:4, if one would like that better. Next we will demonstrate how a march along a portion of the circle of fifths will look, if one uses only pure fifths and octaves.

Let us choose the pitch of C3 with the frequency of f Hz as the starting point. The pitch of G3 then receives the frequency of $3/2 f$ Hz. The fifth of that tone, D4, has the frequency $3/2 f \cdot 3/2 = 9/4 f$ Hz. The frequency of D3 one octave lower will be half as high, or $9/4 f \cdot 1/2 = 9/8 f$ Hz. The fifth above D3 is A3 with the frequency $9/8 f \cdot 3/2 = 27/16 f$ Hz, and the frequency of its fifth, E4, $27/16 f \cdot 3/2 = 81/32 f$ Hz, so for E3 we get $81/32 f \cdot 1/2 = 81/64 f$ Hz. Next fifth is B3 with the frequency $81/64 f \cdot 3/2 = 243/128 f$ Hz. For a diatonic scale it is now sufficient to complement the series of tones already obtained with F3. The tone one octave below has the frequency of $2/3 f$ Hz, and in that case F3 has twice that frequency, or $2 \cdot 2/3 f = 4/3 f$ Hz. Now, we have obtained the diatonic Pythagorean tuning. It is shown in Table 4.1.

The bottom line shows the frequency ratios of the major and minor seconds of this tuning. We can see that all seconds have the same ratio: 9:8 for the major seconds and 256:243 for the minor seconds.

Table 4.1   Pythagorean tuning

	C	D	E	F	G	A	B	C
Frequency	f	$\frac{9}{8}f$	$\frac{81}{64}f$	$\frac{4}{3}f$	$\frac{3}{2}f$	$\frac{27}{16}f$	$\frac{243}{128}f$	$2f$
Frequency ratio	$\frac{9/8}{1} = \frac{9}{8}$	$\frac{81/64}{9/8} = \frac{9}{8}$	$\frac{4/3}{81/64} = \frac{256}{243}$	$\frac{3/2}{4/3} = \frac{9}{8}$	$\frac{27/16}{3/2} = \frac{9}{8}$	$\frac{243/128}{27/16} = \frac{9}{8}$	$\frac{2}{243/128} = \frac{256}{243}$	

If we examine other intervals within the diatonic scale, we see that the fifth, the fourth, and the octave are all just. After some mathematical exercise, we can see that the major third has the ratio of

$$\frac{9}{8} \cdot \frac{9}{8} = \frac{81}{64}$$

This is true regardless if we compute the interval C-E, F-A, or G-B. This third is not identical with the just interval that has the frequency ratio of 5:4 and is found between the fourth and fifth partials in a harmonic spectrum. The interval between the just and the Pythagorean major thirds has the frequency ratio of

$$\frac{5}{4} : \frac{81}{64} = \frac{80}{81}$$

This very narrow interval is called the *syntonic comma*.

The minor third has the ratio of 32:27, regardless if we calculate it as the interval D-F or E-G. This is also different from the just interval, obtained between the fifth and sixth partial in a harmonic spectrum, and having the ratio of 6:5. Also this third differs from the Pythagorean by the syntonic comma of 80:81.

By continuing the march in fifth steps around the circle of fifths, one can compute the chromatic semitone of the Pythagorean tuning. It can be obtained from the upper fifth of the tone B with the frequency of $243/128\,f \cdot 3/2 = 729/256$ f Hz. This is obviously the frequency of the tone F#4, so the tone one octave lower is 729/512. The interval F3-F#3 then becomes 2187/2048. This is the Pythagorean chromatic semitone.

If we continue the trip around the circle of fifths until the 12th fifth is reached above the starting tone, we arrive, after energetic computation including several octave transpositions, at the interval between the starting tone and its 12th fifth. This interval has the magnificent ratio of 531441:524288. It is called the *Pythagorean comma* and can also be found as the interval between the diatonic and chromatic semitone.

VII. JUST TUNING

We just constructed the Pythagorean tuning using only the two intervals between the three lowest partials of a harmonic spectrum: octave and fifth. The just tuning is obtained by using some more of such intervals, those found between the partials up to number 6; one uses the just major triad produced by partials number 4, 5, and 6. As we saw earlier, the just major and minor thirds have the frequency ratios of 5:4 and 6:5, respectively.

Again, we chose the tone C with the frequency of f Hz as our starting point. The tones of the triad C-E-G receive the frequencies of f Hz, $5/4\,f$ Hz, and $3/2\,f$ Hz. As in the Pythagorean tuning, the frequency of the tone F is $4/3\,f$ Hz. Its major third A then receives the frequency of $4/3\,f \cdot 5/4 = 5/3\,f$ Hz. Analogously, the frequency of B, a major third above G with the frequency of $3/2\,f$ Hz gets the frequency of $3/2\,f \cdot 5/4 = 15/8\,f$ Hz and D $3/2\,f \cdot 3/2 = 9/4\,f$ Hz, or, after octave transposition, $9/8\,f$ Hz, just as in the Pythagorean tuning. This completes the building of the just diatonic scale. The construction principle is illustrated in Figure 4.2.

As seen in the figure, we pile up three just triads on top of each other: those of the subdominant (IV), the tonic (I), and the dominant (V). The tuning is presented in its entirety in Table 4.2.

The frequency ratios of the minor and major seconds are shown in the bottom line. We can see that just tuning contains two major seconds of differing sizes. One has the frequency ratio of 9:8, as its Pythagorean cousin, and the other 10:9. Both minor seconds, however, have the same size, 16:15.

The just tuning is similar to the Pythagorean with regard to the fifth, the fourth, and the octave. However, contrary to the Pythagorean tuning, three triads are just, F-A-C, C-E-G, and G-B-D, their frequency ratios being 4:5:6. The minor triads E-G-B and A-C-E are both constructed of just minor and just major thirds (if you doubt that, do the calculation once again). The corresponding frequency ratios can be written 5/4:3/2:15/8, which we can express as 10:12:15. (This, incidentally, tends to charm friends of small integers.)

The greatest serious advantage with the just tuning is that the three major triads contained in the diatonic scale and the two minor triads E-G-B and A-C-E are all just. They do not produce beats as long as harmonic spectra are being used.

The third minor triad in the diatonic scale, D-F-A, is a problem though. The minor third D-F has the frequency ratio 4/3:9/8 = 32:27. This ratio cannot be reduced to a ratio between small integers. Consequently, that third generates beats. The same applies to the frequency ratio of the fifth having the frequency ratio of 5/3:9/8 = 40:27. The just major third F-A cannot help the situation appreciably; this minor triad produces salient beats. This, of course, is a problem.

We saw that just tuning contains two different major seconds, and as a consequence, there are also two minor sevenths of different sizes; the interval D-C (counting upward) has the frequency ratio of 16:9, which is somewhat smaller than the 9:5 of the interval E-D.

These cases in which the same musical intervals appear in different sizes in the scale cause problems in certain situations. Imagine that we tune an organ in

Figure 4.2 Recipe for constructing just tuning.

Table 4.2 Just tuning

	C	D	E	F	G	A	B	C
Frequency	f	$\frac{9}{8}f$	$\frac{5}{4}f$	$\frac{4}{3}f$	$\frac{3}{2}f$	$\frac{5}{3}f$	$\frac{15}{8}f$	$2f$
Frequency ratio	$\frac{9/8}{1}=\frac{9}{8}$	$\frac{5/4}{9/8}=\frac{10}{9}$	$\frac{4/3}{5/4}=\frac{16}{15}$	$\frac{3/2}{4/3}=\frac{9}{8}$	$\frac{5/3}{3/2}=\frac{10}{9}$	$\frac{15/8}{5/3}=\frac{9}{8}$	$\frac{2}{15/8}=\frac{16}{15}$	

just tuning. Assume that we tune the C major scale, i.e., we attribute fixed frequencies to the scale tones. If we then want to play a D major scale, we find that the major seconds of differing sizes appear at the wrong locations in the scale, as illustrated in Table 4.3. In D major, the first second (D-E) has the frequency ratio of 10:9 instead of 9:8, which it should have in just tuning. Similar errors occur in other places. This produces intervals that are far from just in some places. Therefore, the just tuning is inappropriate for playing in different keys on instruments with fixed frequencies. In reality it is almost impossible on keyboard instruments.

It should be pointed out, though, that several attempts have been made over the years to construct keyboard instruments offering the player just tuning. One solution has been to provide the keyboard with more than 12 keys per octave. This, of course, improves the situation to some extent. In recent time, this idea can be realized much more simply if one uses synthesizers supported by some computer machinery. As we will see later in this chapter, in section XIII, certain problems remain also with this solution.

VIII. EQUALLY TEMPERED TUNING

In the two types of tuning presented above, we have noted certain problems that occur when the tuning is implemented on instruments with fixed frequencies for the tones. The Pythagorean is associated with the Pythagorean comma. The just

Table 4.3 Difficulties associated with transpositions within just tuning

Tone	C	D	E	F	G	A	B	C	
Frequency ratio		9:8	10:9	16:15	9:8	10:9	9:8	16:15	
Tone		D	E	F#	G	A	B	C#	D
Frequency ratio is: should be:			10:9 9:8			10:9 9:8	9:8 10:9		

scales have major thirds and other intervals appearing in different sizes within the diatonic scale and also a mistuned triad.

The equally tempered tuning offers a radical solution to all these problems. It is based on the idea that the octave is divided into 12 semitones that are all exactly identical in size. This implies that they all correspond to exactly the same frequency ratio. Another way of saying this is that all fifths along the circle of the fifths are exactly equal in size, and a tiny bit more narrow than just, such that the 12th fifth is exactly a couple of octaves above the starting point. Thus, every fifth is a 12th of a Pythagorean comma too narrow. The derivation of the frequency ratio of the semitone in the equally tempered tuning is a nice demonstration of the beauty of mathematical solutions.

The just octave has the frequency ratio of 2:1. We now want to split this ratio into 12 equally large parts. The simplest way to do this is by counting in terms of *exponents*. Next we will digress for a moment and give a brief presentation for the benefit of those who find exponents exotic.

To raise a number to the power of n (e.g., 3^n) means that the number should be multiplied by itself n times:

$$3^n = \underbrace{3 \cdot 3 \cdot 3 \cdot 3 \cdot \ldots}_{n \text{ items of } 3} \tag{4.13}$$

Thus, if we raise a number to the power of 1 it shall be multiplied with itself 1 time:

$$3^1 = 3$$

Therefore, we can write the frequency ratio of the octave

$$2:1 = 2^1:1^1 \tag{4.14}$$

The splitting into 12 equal parts that we are looking for can be obtained by splitting the exponent into 12 equal parts:

$$2^{1/12}:1$$

which normally is written as

$$\sqrt[12]{2}:1$$

This reads "the 12th root of 2." By multiplying this frequency ratio with itself 12 times, we get

$$\sqrt[12]{2}^{12}:1^{12} = 2^{12/12}:1^{12} = 2:1 \tag{4.15}$$

that is, we retrieve the frequency ratio of the just octave.

Summarizing this little excursion into mathematics, we see that the splitting of the octave into 12 equally wide intervals that we were looking for can be obtained by giving it the frequency ratio of

$$\sqrt[12]{2}:1$$

VIII. Equally Tempered Tuning

This ratio is the frequency ratio for the semitone in equally tempered tuning.

As every semitone in the equally tempered tuning has the same frequency ratio, the frequencies of the scale tones can be computed by means of the frequency ratio of the semitone. The frequency of C#, $f_{C\#}$, should then be related to that of C, f_C as $\sqrt[12]{2}$ is related to 1, or

$$\frac{f_{C\#}}{f_C} = \frac{\sqrt[12]{2}}{1} \quad (4.16a)$$

$$f_{C\#} = \sqrt[12]{2} \cdot f_C \text{ Hz} \quad (4.16b)$$

Similarly, the frequency of the tone D, f_D, should relate to that of C#, as $\sqrt[12]{2}$ is related to 1, or

$$\frac{f_D}{f_{C\#}} = \frac{\sqrt[12]{2}}{1} \quad (4.17a)$$

However, according to Eq. (4.16b), $f_{C\#} = \sqrt[12]{2} \cdot f_C$ Hz
Therefore, we get

$$f_D = \sqrt[12]{2} \cdot \sqrt[12]{2} \cdot f_C = (\sqrt[12]{2})^2 \cdot f_C = 2^{2/12} \cdot f_C = 2^{1/6} \cdot f_C = \sqrt[6]{2} \cdot f_C \text{ Hz} \quad (4.17b)$$

All the other frequencies for the scale tones in the equally tempered tuning can be computed in exactly the same way. These frequencies are shown in Table 4.4. Even if one tries to tidy the expression as much as possible, they may appear as being far from transparent. It is slightly easier to get an idea about the size of the intervals if one expresses the frequency ratios in terms of $1:a$. If we compute the frequency ratio of the semitone in this form, we obtain.

$$1.05946:1$$

in case we content ourselves with five decimals.

This is very worthwhile to remember, because it gives us an idea as to the size of the semitone in the equally tempered tuning. If we are a bit generous, we can round off the above frequency ratio to

$$1.06 = 1 + \frac{6}{100}$$

Table 4.4 Frequency values of tones in the equally tempered tuning scale

	C	C#	D	D#	E	F	F#	G	G#	A	A#	B	C
Frequency	f	$\sqrt[12]{2}f$	$\sqrt[12]{2^2}f$	$\sqrt[12]{2^3}f$	$\sqrt[12]{2^4}f$	$\sqrt[12]{2^5}f$	$\sqrt[12]{2^6}f$	$\sqrt[12]{2^7}f$	$\sqrt[12]{2^8}f$	$\sqrt[12]{2^9}f$	$\sqrt[12]{2^{10}}f$	$\sqrt[12]{2^{11}}f$	$\sqrt[12]{2^{12}}f$
or	f	$\sqrt[12]{2}f$	$\sqrt[6]{2}f$	$\sqrt[4]{2}f$	$\sqrt[3]{2}f$	$\sqrt[12]{2^5}f$	$\sqrt{2}f$	$\sqrt[12]{2^7}f$	$\sqrt[3]{2^2}f$	$\sqrt[4]{2^3}f$	$\sqrt[6]{2^5}f$	$\sqrt[12]{2^{11}}f$	$2f$

Six hundredths is normally called 6%. This means that a semitone is close to a frequency increase of 6%. Thus, a tone with the frequency of 106 Hz is about one semitone higher than one at 100 Hz (or, if we want to be finical, 105.946 Hz).

After energetic mathematical exercises (or after having peeped at an interval table), we can list the frequency ratios for the various intervals in the equally tempered tuning shown in Table 4.5.

The percentages in Table 4.5, which are very coarse, may still be of some help when reading the table. The frequency ratio of the just fifth is 1.5:1, which tells us that 50% frequency increase characterizes a fifth. The frequency ratio of the equally tempered fifth is almost but not exactly 50%. A more exact value is 48.831%, thus slightly less. Later in this chapter we will return to the sizes of this and other intervals and compare them with what they are in just and Pythagorean tuning.

IX. CENT

Above we have seen that the straightforward way of octave splitting used in the equally tempered tuning leads to rather charmless frequency ratios. If we want to compute the frequency ratio for a major third in the Pythagorean tuning, we multiply the frequency ratio of the major second with itself twice: 9/8 · 9/8 = 81/64. If we try the same trick with the frequency ratios of the equally tempered tuning, we face the disgusting task of multiplying the frequency ratio of

Table 4.5 Frequency ratios for intervals in equally tempered tuning (percentages are only approximations).

Interval	Frequency ratio	Frequency increase (%)
Minor second	1:1.05946	6
Major second	1:1.1124	12
Minor third	1:1.1892	19
Major third	1:1.2599	26
Fourth	1:1.33484	33
Augmented fourth / Diminished fifth	1:1.41421	41
Fifth	1:1.49831	50
Minor sixth	1:1.58740	59
Major sixth	1:1.68179	68
Minor seventh	1:1.78180	78
Major seventh	1:1.88775	89
Octave	1:2	100

IX. Cent

the equally tempered minor second (i.e., 1.05946) with itself 4 times. Even with access to paper, pencil, and eraser, this is both a demanding and boring enterprise.

In Chapter 2 we mentioned the advantages of calculating with logarithmic units. This implies that multiplication and division are converted to addition and subtraction. This means that if we express the frequency ratios of the different intervals in logarithmic terms, we would spend a more productive and agreeable life computing musical intervals. The usual unit of this kind is called *cent*.*

One cent corresponds to the frequency ratio of a very tiny interval: a 1,200th of an octave. Thus its frequency ratio is

$$1 \text{ cent} = \sqrt[1200]{2} \tag{4.18}$$

This means that 100 cents is a semitone in the equally tempered tuning. It is worthwhile to store the following in one's long-term memory:

A number of cents corresponds to a frequency ratio.

As cent is a logarithmic unit, multiplication and division correspond to addition and subtraction of cent values. For example, the cent value for a major second in the equally tempered tuning is the sum of the cent values for two minor seconds: 100+100=200 cents. In this simple way we can find the cent values for all intervals in this tuning (see Table 4.6).

Table 4.6 Cent values for intervals in the equally tempered tuning

Interval	Cent value
Minor second	100
Major second	200
Minor third	300
Major third	400
Fourth	500
Augmented fourth / Diminished fifth	600
Fifth	700
Minor sixth	800
Major sixth	900
Minor seventh	1000
Major seventh	1100
Octave	1200

---

*In some French literature, one can find the unit Savart instead of cent; 1 Savart = 3.99 cents. Savart was a French acoustician.

As mentioned, each cent value corresponds to a frequency ratio. As a rule of thumb, a frequency ratio of 1.01:1, i.e., a frequency increase of 1% is equivalent to an interval of 17 cents, and 10 cents corresponds to the frequency ratio of 1.0059846:1, i.e., a frequency rise of about 0.6%. This would appear convincing, as 100 cents, or a minor second in the equally tempered tuning, corresponds to a frequency increase of about 6% (see Table 4.5).

How do we compute the cent value for a given frequency ratio? For small intervals, less than a semitone, one can get a fair approximation from the percentage frequency increase. In such cases one can use the relation just mentioned, that 1% is approximately 17 cents. However, it is very important to observe that percentages cannot be added in the same simple way as cent values. It is only the latter that are logarithmic!

The exact relation between frequency ratio and cent value is as follows. We want to calculate the cent value for an interval with the frequency ratio i. Then, regardless of interval size, that the cent value C for the interval is

$$C = \frac{1200 \cdot \log i}{\log 2} = 3986 \cdot \log i \qquad (4.19)$$

For example, assume that the interval is a fifth (i.e., $i = 3:2$) and $\log i = 0.1761$, which we realize by means of a pocket calculator or a logarithm table. The cent value C for the fifth is then

$$C = 3986 \cdot 0.1761 = 702 \text{ cents}$$

Thus we see that the cent value of the fifth is 702 cents, whereas its cousin in the equally tempered tuning is only 700 cents. Let us now for a moment return to the Pythagorean tuning with its comma. The origin of the comma is that one piles up 12 fifths on top of each other. Each of these is 702 cents. If they were 700 cents, as in the equally tempered tuning, no comma would occur. As $12 \cdot 2 = 24$, we realize that the cent value for the Pythagorean comma must be 24 cents, which is also quite true.

X. COMPARING TUNINGS

The cent unit offers a convenient way to compare the three tunings presented above. Thereby, it is good to know that the smallest perceptible melodic interval is about 5 cents. This is true when the two tones are presented in succession. For dyads, i.e., two tones sounding simultaneously, the smallest perceptible frequency difference is often smaller, because beats rather than a pitch difference can be used as the criterion. Frequencies and cent values for intervals in the various tunings are shown in Table 4.7.

X. Comparing Tunings

Table 4.7   Comparisons between intervals in different tunings

Frequencies	C	D	E	F	G	A	B	C
Extra values								
Pyth.	f	$\frac{9}{8}f$	$\frac{81}{64}f$	$\frac{4}{3}f$	$\frac{3}{2}f$	$\frac{27}{16}f$	$\frac{243}{128}f$	$2f$
Just	f	$\frac{9}{8}f$	$\frac{5}{4}f$	$\frac{4}{3}f$	$\frac{3}{2}f$	$\frac{5}{3}f$	$\frac{15}{8}f$	$2f$
ET	f	$\sqrt[6]{2}f$	$\sqrt[3]{2}f$	$\sqrt[12]{2^5}f$	$\sqrt[12]{2^7}f$	$\sqrt[4]{2^3}f$	$\sqrt[12]{2^{11}}f$	$2f$
Approximate values								
Pyth	f	$1.13f$	$1.27f$	$1.333f$	$1.5f$	$1.69f$	$1.90f$	$2f$
Just	f	$1.13f$	$1.25f$	$1.333f$	$1.5f$	$1.67f$	$1.88f$	$2f$
ET	f	$1.12f$	$1.26f$	$1.335f$	$1.498f$	$1.68f$	$1.89f$	$2f$

Cent values for some intervals

Interval	Pyth.	Just	ET
Minor second	90	112	100
Major second	204	{ 182 / 204 }	200
Minor third	294	316	300
Major third	408	386	400
Fourth	498	498	500
Augmented fourth	590	590	600
Diminished fifth	588	610	600
Fifth	702	702	700
Minor sixth	792	814	800
Major sixth	906	884	900
Minor seventh	996	{ 996 / 1018 }	1000
Major seventh	1110	1088	1100
Octave	1200	1200	1200

Note that for each interval the cent values of the equally tempered tuning are either very close to or between those of Pythagorean and just. One could say that the equally tempered tuning serves as a compromise between these two tunings. We can also observe that the difference between Pythagorean and just is 22 cents for all intervals where they differ. Note also that the just major third is narrow and the just minor third is wide as compared with Pythagorean tuning. This fact is often not observed; a common view is that the just major third is wide and the just minor third is narrow.

The tunings presented above are by no means universal, although they seem to be pretty relevant to our music practice, as we will see soon. However, many other tunings apart from those presented above have been used. To widen our views

Table 4.8 Cent values for Schlick's mean tuning temperament

	C	C#	D	D#	E	F	F#	G	G#	A	A#	B	C
Cent value	0	90	196	302	392	502	590	698	796	894	1002	1090	1200

somewhat, Table 4.8 shows an example, Schlick's mean tuning temperament, which is frequently used for playing old music on authentic keyboard instruments. Shown are the intervals between the reference tone C and the tones of the chromatic scale.

Also in this case we note that the intervals of the tempered tuning fall between their Pythagorean and just cousins. It is also interesting that the fifth is no less than 4 cents narrower than just; however, the 2-cent difference between the fifths of the equally tempered and the Schlick tuning is too small to be perceptible as a melodic interval by normal listeners. But if presented as a dyad the difference would be easy to detect.

There are also many other types of tunings; as soon as we reach outside the framework of our Western art music, quite different tunings and scales are encountered. To counteract the imminent ethnocentricity somewhat, Table 4.9 presents measurements of tunings for two tuning systems from the Indonesian cultural sphere: the seven-tone Pelog and the five-tone Slendro.

Evidently, there is not very much of a similarity between this scale and the ones we have seen in this chapter. The Pelog lacks most consonant intervals when counted with tone number 1 as the reference. Further, the interval patterns are not rigid but vary between players and sometimes also to some extent between octaves in the same tuning.

Perhaps, our Western intervals are optimized for our instrument families and the ways in which they are used. For example, the equally tempered tuning produces rather slow beats in consonant dyads. One might suspect that they optimally fit polyphonic music, which is realized with harmonic spectra and in which consonance and dissonance play a prominent role in the grouping of note sequences into blocks, phrases, etc.

Table 4.9 Cent values for two Indonesian tunings

	Tone	1	2	3	4	5	6	7	1
7-tone Pelog			250	120	150	270	150	115	165
Interval to tone 1			250	370	520	790	940	1055	1220
	Tone	1	2	3	4	5	1		
5-tone Slendro			240	240	240	240	240		
			240	260	255	255	190		
Interval to tone 1			240	480	720	960	1200		
			240	500	755	1010	1200		

XI. CALCULATING INTERVALS

Imagine that we have gathered a series of frequency values from some kind of music, and we are now curious to find out what the intervals are between the notes.

The best way is to get access to a pocket calculator. First, determine the frequency ratio, which, for instance, will be 1.5 for a fifth. Then, press the button marked "lg" (or "log" on some calculators), which gives a number series starting with 0.17609 . . . . Finally, multiply this value with the constant mentioned before, 3,986, which applies as long as we use the logarithm with 10 as the base. We obtain 701.899, or 702, for short. (As we might recall from Chapter 2, there are other bases also. There is no harm in using these, but then, the constant will not be 3,986 any more.)

If one does not like pocket calculators, a logarithm table will also do, even though it takes more time and does not offer the same accuracy. We just saw that the cent value of an interval with the frequency ratio of i is $3,986 \cdot \log i$. As the use of logarithms convert multiplication to addition and division to subtraction, we can compute the cent value C as

$$C = 3986 \cdot (\log f - \log g) \qquad (4.20)$$

where f and g are the frequencies of the tones constituting the interval. Let us now assume that $f = 428$ Hz and $g = 316$ Hz. The logarithm table then informs us that

$$\log 316 = 2.4997 \quad \text{and} \quad \log 428 = 2.6314$$

To arrive at nice positive numbers, we always subtract the smaller logarithm from the greater, rather than the reverse. Then we get

$$C = 3986 \cdot (2.6314 - 2.4997) = 3986 \cdot 0.1317 = 525$$

There are also some shortcuts available, which sometimes may be useful. Every musician and music acoustician knows the harmonic overtone series and the intervals between the 10 lower partials in a harmonic spectrum. Then, one automatically has access to the frequency ratios of the main intervals according to just tuning. If we disregard the subtleties of tunings, these would do reasonably well as approximations for the corresponding intervals. If, in addition, the frequencies of two tones of interest can be written as a ratio between small integers (smaller than 10), one can venture an intelligent guess as to what the interval between these interesting tones should be.

Sometimes one is so lucky that the ratio between the frequencies can simply be reduced. For example, if one frequency is 85 Hz and the other is 119, it is actually possible to reduce the ratio by dividing numerator and denominator by 17; 85/17 = 5; 119/17 = 7. Thus, the frequency ratio is 5:7, and the interval will be the same as between partials number 5 and 7 in a harmonic spectrum. This interval is close to a diminished fifth.

It is, of course, rare that such simple reductions are possible. Sometimes it is enough to simply add some very small quantity to the numerator or the denominator to obtain a ratio that can be reduced. For instance, by being a bit rude, one can say that 85:120 is rather similar to 85:119, and in that case the interval is close to the 5:7 we had before. However, this method is recommendable only for very rough estimates, and investing in a pocket calculator possessing the log function is absolutely much wiser.

In cases where one has long tables of pairs of frequency values that should be converted to intervals in cents, the only acceptable solution in our time is to get hold of some computer provided with a calculation program. Computers have a fabulous ability to repeatedly apply the same equation many times to new values, and often it is very simple to implement the equation that should be applied. It is sometimes argued that computers have a spectacular ability of doing things fast and wrong. Therefore, it is mandatory to make sure that the processing of a series of numerical values is correct. This is done by computing samples by hand or by means of a pocket calculator.

Above we have talked about two kinds of logarithmic entities, cents and decibels. By chance there is an obscure relation between them, which is quite helpful in cases when one wants to find out the percentage corresponding to a certain number of decibels. However, it is necessary to keep one's feet firmly on the ground and one's mathematical head perfectly cool. Readers doubting their ability to do so should immediately skip the rest of this theoretically very violent paragraph and start with the next! The secret is that a doubling of frequency is equivalent to 12 semitones and a doubling of sound pressure is equivalent to 6 dB. Thus, doubling produces twice as many semitones as decibels. If the question is what pressure amplitude ratio corresponds to a certain number of decibels, the intervals between the partials in a harmonic spectrum may help! Suppose we want to find out the pressure ratio for 4 dB. Turning to frequencies these 4 dB would correspond to 8 semitones, and 8 semitones is the interval of a minor sixth. The frequency ratio of this interval is 3:5, or 0.6:1. A sound pressure difference of 4 dB therefore corresponds to a pressure ratio of approximately 6:10. However nice this little shortcut between frequencies and pressure amplitudes may seem, it should not lead us to confuse pressure and frequency!

XII. DEMANDS ON A SCALE

Why is the scale divided into 12 semitones and why do we use major and minor scales at all? The answers are probably tied to the historical development of our music culture. We live in a music tradition in which several instruments usually play simultaneously. This means that our music contains many dyads and chords. We know that consonant dyads produce beats if the intervals deviate from the just

XII. Demands on Scale

values and are realized on instruments producing harmonic spectra. This last condition is fulfilled on most traditional instruments, and beats appear to be disliked in many music cultures. This has probably directed evolution toward scales containing several beat-free intervals, such as octave and fifth.

Consonance and dissonance represent a timbral dimension that the composers in our Western cultural tradition have certainly been making frequent use of. If we think about it thoroughly, it is this frequent use that has lead to the theory of harmonic progressions, which has been so efficiently used for marking musical structure in the past; for example, the classical way of signaling the end of a structural unit such as phrase or subphrase is by means of a *cadence,* a special pattern of chord progressions. If one wants to use the timbral dimension consonance/dissonance in composition, one needs a scale offering dyads ranging from very consonant to very dissonant. The octave represents the highest possible degree of consonance that can be achieved with harmonic spectra and thereafter comes the fifth and the major and minor thirds, whereas the minor second is the most dissonant interval possible and then comes its major cousin. Thus a great demand for a wide range of variation with regard to dissonance/consonance seems a good candidate for explaining the development of the diatonic scale.

When keyboard instruments with fixed frequencies became common, another demand appeared: Transposition should be possible without ruining the tuning of the dyads between scale tones. The equally tempered tuning represents the optimal solution to this problem. It contains departures from just intervals that are so small that the beats are still bearable, at least if one does not deliberately listen for them.

If our diatonic scale has emerged under the pressure from the harmonic spectrum, from the great significance attributed to variation in consonance/dissonance of dyads, and from the need for transposition, one would expect to find other types of scales in other cultures developed under different conditions. This is also the case. If one turns to the ethnic music cultures that always existed in parallel with the main cultural tradition with its more sophisticated and thoroughly organized ensemble music, one finds scales that do not even contain fifths or major or minor thirds, but rather something close to an augmented fourth or a perfect compromise between a minor and a major third. One important factor may be that in such music cultures the tones are played one at a time rather than in dyads or chords. Another significant factor may derive from the practice used in hole drilling in the manufacturing of flutes.

Most instruments in our music culture produce harmonic spectra, as mentioned. However, in the contemporary computer-aided electroacoustic music studios, this is not a necessary constraint any longer. One would then ask if this does not open up quite new possibilities also with respect to harmony. If one decides to use one particular kind of inharmonic spectra for all tones, it should be possible to tailor a new scale and a new harmony to this inharmonicity. An interesting attempt has been done along these lines by John Pierce and Max Mathews of Stanford University in California. This attempt is extensively presented in several articles. One of them can be found, including sound examples with compositions applying the new scale and harmony, in the book *Harmony and Tonality,* published by the Swedish Academy of Music.

We saw in the preceding chapter that the nonlinearity of the human hearing system generates harmonic spectra, as soon as two sine tones excite the ear. In a sense, the ear seems to have a natural disposition for harmonic spectra. An interesting question therefore is whether this inclination for harmonic spectra and just intervals is so influential that playing with inharmonic spectra is an unrealistic idea from the outset. The experiments just mentioned seem to suggest that the prejudice of the ear is reasonably insignificant.

XIII. THE SCALE IN PRACTICE

Do tuners and musicians obey the theories for scale tone frequencies that we have presented above, or do the theoretical tunings deviate substantially from those used in sounding music? This question has not been investigated to any great extent yet, but there are some relevant studies to mention. Because the vibrato eliminates the risk for beats in mistuned consonances, it is an important factor, allowing the musicians a certain degree of freedom.

Considerable care is spent on tuning in so-called barbershop singing, which is performed without vibrato by a quartet of male singers. Table 4.10 shows mean values for the tuning of the tonic and dominant seventh chords in two very successful ensembles in this genre.

Under these conditions, neither the narrower just or the wider Pythagorean major third were used, but rather something more similar to an equally tempered version. Further, the fifth was somewhat wider than just. The seventh represented

Table 4.10 Mean sizes and confidence intervals for dyads in tonic (T) and dominant seventh (D) chords as performed by two barbershop quartets, repeatedly rendering a cadence a great number of times. (Measurements by Hagerman and Sundberg, 1980.)

Dyad	Context	Size (cent)	Confidence interval (cent)
Major third	T:I&III	403	400–406
Major third	D:I&III	396	391–400
Fourth	T:V&I	493	491–496
Fifth	T:I&V	705	702–709
Minor seventh	D:I&VII	977	971–983
Octave	T:I&I	1199	1196–1201

XIII. The Scale in Practice

a particularly interesting case. In the dominant seventh chord, it was tuned very narrow, 23 cents below equally tempered. This is in the vicinity of the minor seventh contained in the harmonic series between partials number 4 and 7, thus having the frequency ratio 4:7. This interval is no less than 33 cents narrower than equally tempered. Under vibrato-free conditions, just tuning thus seems relevant.

Shackford (1961 and 1962) investigated the tuning of dyads and chords as well as melodic intervals in ensemble playing. Here, the musicians played with vibrato, so the risk for beats was avoided. The results revealed a rather considerable variability that could, at least in part, be ascribed to the musical context. Some results are shown in Table 4.11.

There were no dramatic differences between simultaneous and successive intervals with respect to major second and fourth. A small difference can be seen for the fifth, the dyad version being close to the barbershop result of 705 cents. One can also see that these musicians preferred the Pythagorean major third to the just. It appears that the intervals are tuned according to just, equally tempered, and Pythagorean tuning depending on the musical context.

The differences between these tunings are not all that great after all. Are they still large enough to be relevant? Table 4.12 shows mean values of settings from an investigation in which 17 musically experienced subjects were asked to adjust the frequency of a variable synthetic sung tone so that it constituted certain dyads with another synthetic sung tone. Both tones had vibrato and each subject made 10 settings of each of the 4 intervals. The "elite" was a group of five subjects who showed the highest consistency of their settings.

The values show that there is a reasonable agreement between the ideas of the ideal fifth and major third according to these listeners and the musicians who produced the values shown in Table 4.11. This is, of course, not unexpected. After all, musicians play for listeners.

Table 4.11 Mean values (MV) and quartile deviations (QD) of interval sizes measured in a string trio as compared with just, equally tempered, and Pythagorean tunings (J, ET, and P) according to Shackford (1961 and 1962)

Interval	Dyads		Melodic		Closest tuning (cent)
	MV (cent)	QD (cent)	MV (cent)	QD (Cent)	
Minor second			93	86–101	90 (P)
Major second	204	197–211	204	199–209	204 (P)
Minor third	305	287–318			315 (J)
Major third	410	402–418			408 (P)
Fourth			501	408–510	500 (ET)
Fifth	707	699–714	701	692–708	702 (P&J)

Table 4.12 Average sizes and standard deviations (SD) for dyads between vibrato tones adjusted by musically experienced listeners. Elite shows the results from the 5 most consistent subjects. (Data from Ågren, 1976.)

		Average SD	
Interval	Size (cent)	All subjects (cent)	Elite (cent)
Major second	199	14	9
Major third	402	9	6
Fifth	704	10	6
Octave	1204	10	6

It is interesting that the most consistent subjects show very small standard deviations. These are, in fact, very close to the DL for frequency. It also seems that listeners were more meticulous with the consonant dyads than with the more dissonant major second, despite the fact that there were no beats to use as a criterion. Anyway, these results indicate that the demands on accuracy in tuning are very severe in music practice. Even error of 6 cents seems to be a bit on the large side for a fastidious music listener. If expressed in frequency, 6 cents is no more than 0.3 Hz at 100 Hz! And the conclusion regarding just, equally tempered, and Pythagorean tuning is evident: They clearly matter not only in theory but also in reality.

From some other studies of intonation in practice, a clear indication of tuning differences between dyads and melodic intervals emerged. Under vibrato-free conditions, subjects tended to tune dyads in the vicinity of the just version, as we saw. However, if melodic intervals are tuned according to the same recipe, the result sounds horribly out of tune. Values for various melodic intervals according to adjustments performed by a professional violin player are shown in Figure 4.3. The horizontal axis shows the number of fifths contained in the interval, counting ascending fifths positive; further, all intervals are related to the tone C, and the values show the tuning as compared with the equally tempered models. Thus, the major second C-D, containing two fifth steps (C-G, G-D), appears at the value of 2, and the minor second C-D♭ is found at −7.

It is possible that the basic principle underlying most of the results shown in Figure 4.3 reflects the desire to obtain narrow minor seconds. Table. 4.11 showed that this interval was performed clearly narrower than equally tempered. In traditional tonal music, a flattened tone is frequently followed by a descending minor second, and a sharpened tone is often followed by an ascending minor second. For example, the interval C-E♭ would be followed by the tone D more often than by the tone E. By tuning this minor third narrow, E♭ approaches D, and hence this

XIV. Craving for Stretching

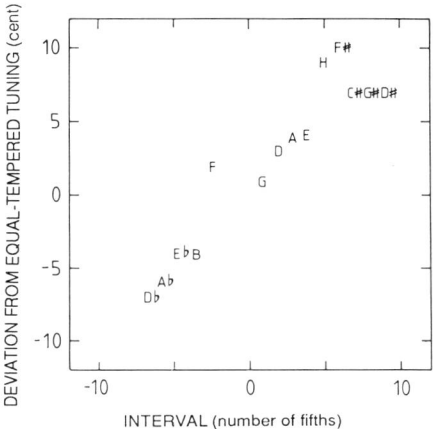

Figure 4.3 Mean deviations from the equally tempered tuning of melodic intervals adjusted by a professional violinist. The horizontal axis shows the number of fifths contained in the interval, counting ascending fifths positive and descending fifths negative. All intervals are related to tone C, and the symbols represent target tones.

minor second comes out narrow. Conversely, in the minor third C-D♯, the latter tone is sharpened, so that the minor second D♯-E becomes narrow. However, this demand for narrow minor seconds does not explain why the number of fifths contained in the interval is relevant.

It is evident from Figure 4.3 that the higher the number of *ascending* fifths contained in the interval, the wider the interval tends to be played, and the higher the number of *descending* fifths there is, the narrower. It is interesting to recall that Pythagorean tuning implies the use of just fifths, which are 2 cents wider than equally tempered. Thus, if Pythagorean values had been used, the deviation from the equally tempered tunings would increase by 2 cents per ascending fifth and decrease by −2 cents per descending fifth. The figure shows that this musician tuned these melodic intervals somewhere between the Pythagorean and the equally tempered. Expressed in Pythagorean terms, the fifths underlying the values in the figure would correspond to about 701.5 cents on the average.

XIV. CRAVING FOR STRETCHING

It is quite remarkable that musicians seem to prefer too wide or "stretched" intervals in many cases. Above we have seen several examples of interval stretching: the barbershop singers' fifth and just minor seventh; string trio players' melodic major and minor thirds and fifths; music listeners preferred sizes of fifth and octave; and a professional musician's settings of melodic intervals that contain ascending fifths.

In the case of octave, the craving for stretching has been noticed for both dyads and melodic intervals. The amount of stretching preferred depends on the mid frequency of the interval, among other things. The average for synthetic, vibrato-free octave tones has been found to be about 15 cents. Thus, subjects found a just octave too flat but an octave of 1,215-cent just. If it had not been for the prevalence of harmonic spectra, we could have expected that the theoretically just octave would have had to yield for the slightly wider, perceptually just one.

An interesting experiment concerning the expressive aspects of octave tuning was performed by the psychologist Scott Makeig. By filtering speech, he eliminated everything but the voice pitch from the speech signal. Then he presented the remaining sound several times, every second time transposed up by one octave. However, the exact amount of transposition was varied slightly, such that stretched, just, and flat octave transpositions were tried. The subjects were asked to judge how the octave-transposed repetition sounded. It turned out that the subjects thought the repetition sounded "certain" and "affirmative" when the octave was stretched and somewhat "insecure" or "doubting" when the octave was just or flat. This suggests that tuning may be a useful tool for musical expression, at least in solo playing.

This idea receives some support also from the results shown in Figure 4.3. It showed that the tuning was influenced by the position of the target note along the circle of fifths. There are reasons to assume that intervals containing several fifths are more "melodically charged" than intervals containing fewer fifths. If so, the stretch can be seen as an expressive means, which the player uses to demonstrate this charge. In a similar way, the tone one octave up may under certain conditions appear as more charged than its colleague one octave below; if this is true, the stretching might be made to mark this difference in charge.

As will be explained in more detail in Chapter 6, the partials of the piano are not quite harmonic. The separation between neighbor partials is somewhat greater than in a perfectly harmonic spectrum. In fact, this leads to stretched octaves in piano tuning. This is a necessary consequence of the inharmonicity and the demand for minimizing beats. The second partial, which is a tiny bit higher than a just octave, serves as the reference when the tone one octave above is being tuned. This obviously leads to stretched octaves. However, as we have just seen, listeners prefer stretched octaves to just octaves. In a way, then, the piano seems to be the fruit of a fascinating coincidence. The string characteristics enforce stretched octaves, and the ear happens to like such octaves! Would the piano be a musically possible instrument if that had not been the situation? Probably it would still be useful, because other keyboard instruments are tuned with theoretically just octaves, such as harpsichord and pipe organ. But the music composed for the piano would probably be written in a different fashion.

XIV. Craving for Stretching

A German acoustician, Ernst Terhardt, developed an interesting theory that proposes an explanation for why we are so eager to stretch octaves. He departed from the fact that the pitch of a sine tone is changed if another sine tone starts to sound simultaneously. The net result is that both tones push the pitch of the other tone away from its own pitch to increase the distance between the two. In this way, the pitch interval between the first two partials in a harmonic spectrum becomes a bit stretched. Already the unborn fetus is thus exposed to brainwashing when perceiving the sound of its mother's voice, which is constituted by harmonic partials. Terhardt believes that this stretched octave follows us from cradle to casket and that it is this octave that musicians model when they play.

CHAPTER 5

Wind Instruments

I. INTRODUCTION

The common feature of all music instruments is that they convert energy, supplied in one form or the other, to acoustic energy, i.e., a regularly oscillating sound pressure. Music instruments can be grouped according to the principles used in this conversion.

In wind instruments it is a steady airstream that is converted into a pulsating one. If you are familiar with electrical terms, you might prefer to think of these instruments as some kind of device that transduces a DC to an AC signal. All instruments that use an airstream for the sound generation are called *wind instruments*.

The human singing voice is certainly the most common member of the wind instrument family. (The traditional separation of music and singing is illogical from most points of view.*) In the voice, an airflow from the lungs is converted to sound. The wind instrument family also contains instruments that are blown by a player, e.g., the oboe or the trumpet, and instruments using fan and bellows, such as the pipe organ, or containing blown reeds, e.g., harmonium, accordion, or mouth organ.

The wind instrument family is rather great. This simply means that it is very common in our music culture to use airstreams for tone generation.

The conversion of airflow to sound is the common feature of wind instruments. However, the conversion happens in different ways in different types of instruments. In most instruments there is an *oscillator*. It works like a kind of supply valve that quickly oscillates between open and closed. The airflow is therefore arrested at regular time intervals by this valve, so that the input airflow is converted

---

*For instance, the voice functions in a way very similar to that used in an old exotic keyboard instrument called *regal* and its organ stop relative with the same name. Here, the sound is generated in pipes with long reeds and short resonators. For these reasons, the distinction between music and singing should not be supported by any thinking individual.

1. Introduction

to a sequence of air pulses, i.e., to a pulsating airflow. Examples of such oscillating valves are the vocal folds, the brass instrument players' lips, and the reed of the clarinet.

In some other wind instruments the oscillator function is provided for in a different way. The airstream is given the form of a blade, which is brought to oscillation. This gives rise to a pulsating airflow through the instrument. All reed and flute instruments offer examples of this principle.

A pulsating airflow generates a complex tone consisting of a series of harmonic partials (i.e., a harmonic spectrum). This tone is more or less dependent on the resonator that is included in the instrument. All wind instruments contain a tube resonator of some kind. Its resonance frequencies and bandwidths have a quite decisive significance for the oscillator function in a majority of wind instruments.

However, there is a group of instruments in which this is not the case. The most common member of this group is the human singing voice, in which the vocal fold vibrations are practically altogether independent of the resonator. Also in this group of instruments the resonator plays a very important role, namely, as a filter. During the travel through the tube resonator, the partials are attenuated to different extents, depending on their distance to the resonance frequencies. If the frequency of a partial coincides with that of a resonance, this partial is relatively strong, and if it is far away from its closest resonance, it will be considerably weaker.

The fundamental frequency is determined by two different principles in wind instruments. In those instruments in which the sound generation is independent of the resonator, the resonator is short in the sense that its first resonance frequency is much higher than the fundamental frequency of the oscillator. Examples of instruments using this principle is the voice, the regal, and the regal stops of the organ. In these instruments the fundamental frequency is determined by the properties of the oscillator itself: the tension and vibrating mass of the vocal folds, or the length of the vibrating part of the reed. In all other wind instruments the resonator is long in the sense that its lowest resonance frequency is near or below the fundamental frequency. This frequency equals a resonance frequency of the resonator, the lowest one or some of the lower ones.

In these instruments in which the resonator determines the fundamental frequency, an effect called *feedback* is used. Feedback implies that some portion of the thing produced by a device is used for controlling the production process in the same device. In instruments using feedback, a rather big portion of the sound pressure oscillation produced by the oscillating valve function is used for controlling the valve's frequency of oscillation. Therefore instruments in which feedback is used for this purpose can be called *feedback instruments*. The remaining wind instruments working without this feedback can be called *no-feedback instruments*.

Among the feedback wind instruments there are three different kinds of oscillator valves. One type, used in, for example, clarinets and bassoons, uses a pair of thin cane reeds, which are brought to vibration; thereby they alternately open and close the air passage into the resonator. These instruments can be called *cane reed instruments*. The instruments in which the player's lips serve the corresponding purpose are mostly called *lip reed instruments;* however, they are often also simply called *brass instruments*. This is somewhat odd because far from all instruments in that family are made of brass. The third group, in which an airstream is shaped as a blade that is brought to oscillation, can correspondingly be called *air reed instruments*.

After this survey of different types of wind instruments, it might be appropriate to describe in more detail the function and acoustic characteristics of the different instrument groups. However, the most decisive component is the resonator, so first we will spend some time presenting tube resonators of various types.

II. TUBE RESONATORS

The resonances of tube resonators are highly relevant to the spectrum and in most of the wind instruments also to the fundamental frequency. Here we will examine how these very important aspects of tone depend on the length and shape of the tube resonator.

If one wants to measure the resonance frequencies and bandwidths in a resonator, several different methods can be used. One is that a tone with constant amplitude is supplied in one end, sweeping slowly from bass to treble, while in the opposite end a microphone is recording the resulting pressure. Figure 5.1 offers an example.

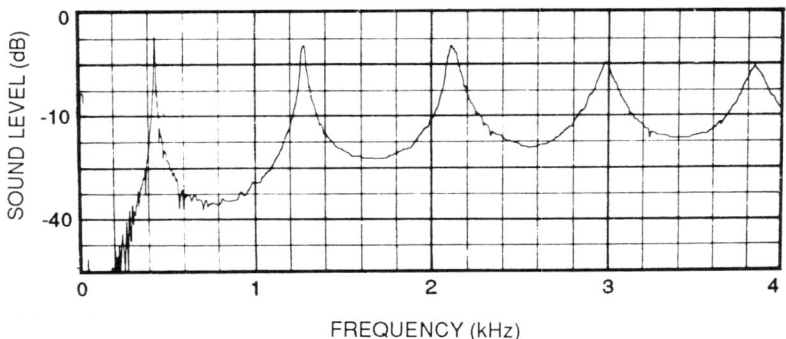

Figure 5.1 Resonance curve for a resonator obtained when a sine tone of constant amplitude, sweeping slowly from bass to treble, is fed to one end. The resulting sound pressure measured in the opposite end, is recorded above as function of frequency. Under these conditions, a resonance appears as a peak in the frequency curve. The sharpness of this peak reflects the losses in the resonator and can be measured in terms of its bandwidth.

II. Tube Resonators

All wind instruments contain a resonator that is mostly made up of an air column enclosed in a tube. This tube may be open in both ends, as in the flute (the embouchure acts as an opening), or only in one end, as in the oboe, clarinet, etc. (One end is practically closed by the reed). The tube can be entirely or partly cylindrical, conical, or horn-shaped. A cylinder possesses a constant diameter, whereas in a cone the diameter increases linearly along the length axis. In a horn, it increases according to a more complicated function. The resonance properties are entirely determined by the exact shape of the tube.

The reason why air columns enclosed in tubes act as resonators is that (1) the air possesses mass and compliance (see Chapter 2,XII) and (2) the sound is reflected at the tube ends, regardless if closed or open. The shape of tube resonators varies a lot. Figure 5.2 shows how the amplitude of air particle velocity is distributed along the length axis of the tube for the first three standing waves (see Chapter 2) in a cylindrical resonator closed in both ends. This is a simple and instructive example of a tube resonator, and therefore we will consider it first, even though it may seem a bit boring from a practical point of view; there will be small success in attempts to emit sound through closed tube endings!

The sound wave is reflected at the closed ends, as a ball hitting a wall. It is not difficult to realize that the air particles next to the ends must be zero: We all know that a hard wall stops motion. If the wave length is such that a velocity node appears simultaneously in both ends, direct and reflected waves are co-operating in phase. This means that resonance occurs. At both closed ends the sound pressure oscillates at a maximum amplitude between over- and under-pressure while the velocity of the air particles next to the end remain still all the

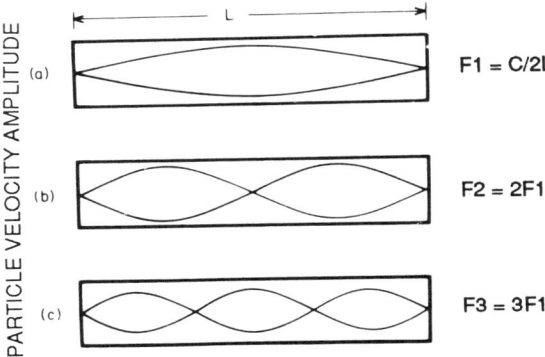

Figure 5.2 Standing waves for the three lowest resonance frequencies, shown in terms of the air particle velocity amplitude, in a tube closed in both ends and with the length of L. At both ends the velocity remains zero.

time. In other words, in a standing wave the sound pressure and the particle velocity are in counterphase.

It is not really difficult to grasp the truth of this. Imagine how the water behaves in the bathtub. If one wants it to surge, one can press down the surface near the end of the tub, i.e., raise the pressure at constant time intervals. In the middle, however, it is appropriate to push the water back and forth along the length axis of the tub, i.e., to give it velocity.

In the ideal case a standing wave can occur at all those frequencies with a wave length (λ) such that exactly one, two, or more (n) half wavelengths can be accommodated in the tube length (l):

$$n \cdot \frac{\lambda}{2} = l \quad (n = 1, 2, 3, 4, \ldots) \tag{5.1}$$

As a consequence of this relation between the tube length and half of the wavelength ($\lambda/2$), tubes with both ends closed are called *half-wave resonators*. As the wavelength equals the ratio between the speed of sound propagation c and the frequency f [see Eq. (2.2)]

$$\lambda = \frac{c}{f}$$

the resonance frequencies in such a pipe occur in the vicinity of

$$n \cdot \frac{\frac{c}{f}}{2} = l; \quad n \cdot \frac{c}{2f} = l; \quad f = \frac{nc}{2l} \quad (n = 1, 2, 3, 4, \ldots) \tag{5.2}$$

Assume the tube length is 1.7 m and the speed of sound propagation c is 340 m/sec. Then,

$$\frac{c}{2 \cdot l} = \frac{340}{2 \cdot 1.70} = \frac{340}{3.40} = 100$$

so the first three resonances occur at

$$1 \cdot 100 = 100 \text{ Hz}$$
$$2 \cdot 100 = 200 \text{ Hz}$$
$$3 \cdot 100 = 300 \text{ Hz, and so on.}$$

The resonance frequencies appear to constitute a harmonic series, but this is not true in reality. The frequency distance between them is successively reduced, the

II. Tube Resonators

higher the number of the resonances. This is due to viscosity* losses along the tube walls.

A different type of resonator, being of greater practical importance, is a cylindrical tube open in both ends. We saw in Chapter 2,XII that a wave is reflected also in an open end. In the ideal case, a tube resonator with both ends open therefore has the same set of resonance frequencies as an equally long tube closed in both ends (see Figure 5.3).

A difference is that pressure oscillations of maximum amplitude do not fit into an open end. Instead, resonance occurs at frequencies with wavelengths such that velocity amplitude antinodes occur at both ends. In this case the air particles rush in and out of the open ends with a maximum amplitude at resonance, while the pressure oscillations are small.

Another difference is that the deviations of the resonance frequencies from a harmonic series (i.e., their inharmonicity) are greater as compared with a similar closed-closed tube. The reason for this is that the sound reflection takes place differently in closed and open ends. However, these differences are not substantial enough to prevent the open-open tube resonator to be of the same half-wavelength type as the closed-closed tube resonator. Flutes of different types contain such half-wavelength resonators, because the embouchure and the mouth act as apertures. When overblown, the pitch jumps one octave up, because the next resonance

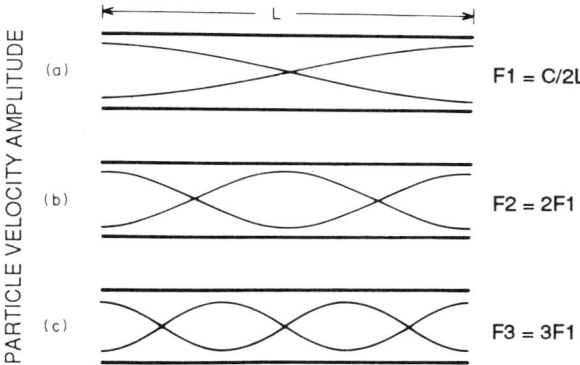

Figure 5.3 Standing waves of the three lowest resonances, represented in terms of the amplitude of the air particle velocity, in an open-open tube of length L. For all resonances the velocity amplitude is maximum at the ends.

---
*Viscosity losses occur because the mobility of the air is not perfectly the same along the tube walls as farther toward the center; the rubbing along the walls slightly reduces the motion of the air particles.

that can be used for controlling the fundamental frequency is one octave higher. What, exactly, overblowing stands for will be explained a bit later.

The inharmonicity of the resonances can be accounted for in terms of a *length correction*. It refers to the mentioned phase characteristics of the reflection but may be easier to imagine. The shape of the air column keeps the form of a rod even after the tube ends. Therefore, the air column is slightly longer than the tube. As a result, the exact location of the reflection is not in the aperture, but a tiny bit beyond. The length correction is the distance between this imagined location of reflection and the real end of the tube. The correction is large at low frequencies and smaller at higher frequencies. This frequency dependence of the length correction explains the inharmonicity of the resonance frequencies.

Given the situation in the two types of resonators already discussed, it is easy to imagine where the resonances will appear in a cylindrical tube open in one end and closed at the other, an open-closed tube, (Figure 5.4). Here, direct and reflected waves cooperate, when the wavelength is such that there is a velocity node at the closed end and a velocity antinode at the open end. The shortest distance between a node and an antinode on a standing wave is a quarter of the wavelength. A resonator of this type is called a *quarter-wave resonator*. In the ideal case resonance occurs at all frequencies with wavelengths such that an odd number $(2n-1)$ of resonances of a quarter of the wavelength equals the pipe length (l). In that case, the condition of a node and an antinode at the ends is fulfilled.

Expressed as a formula the ideal case looks as follows:

$$(2n - 1)\frac{\lambda}{4} = l \qquad (n = 1, 2, 3, 4, \ldots) \tag{5.3}$$

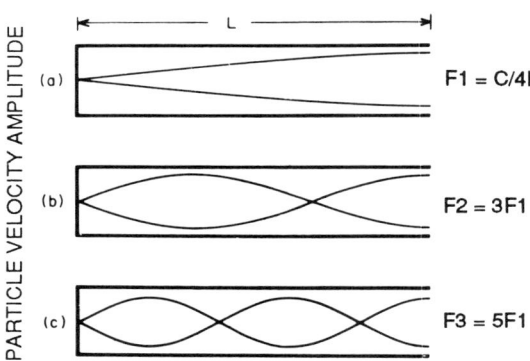

Figure 5.4 Standing waves of the three lowest resonances, represented in terms of the amplitude of the air particle velocity, in an open-closed tube of length L. For all resonances the velocity amplitude is zero at the closed end and maximum at the open end.

II. Tube Resonators

For example, assume that the tube length $l = 1.7$ m and $c = 340$ m/sec., the first resonance appears at

$$\frac{(2 \cdot 1 - 1)340}{4 \cdot 1.70} = \frac{340}{6.80} = 50 \text{ Hz}$$

The second resonance frequency

$$\frac{(2 \cdot 2 - 1)340}{4 \cdot 1.70} = \frac{1020}{6.80} = 150 \text{ Hz}$$

(i.e., three times the frequency of the first one). Similarly, the third is five times higher and fourth seven times higher. We also note that the lowest resonance is one octave below the lowest resonance in an open-open or closed-closed tube of equal length.

The resonance frequencies in this type of resonator are somewhat inharmonic, as in the previously discussed resonators. A music instrument using a cylindrical open-closed tube as the resonator is the clarinet. When overblown, the pitch rises by a duodecime (octave plus fifth), because the frequency of the second resonance is three times that of the first, as we just saw.

For all tube-resonator instruments, it applies that the resonance frequencies are very sensitive to the size of the aperture. Thus, if the aperture is narrowed, the resonance frequencies drop, and vice versa. In many organ pipes this is used as the tuning device.

Up to now we have studied cylindrical tube resonators with open or closed endings. In many instruments, conical tubes are used instead, where the inner diameter increases linearly with the distance to the embouchure end. There is no simple explanation why conical tubes differ from cylindrical tubes, so the best is to simply accept this as a matter of fact. Figure 5.5 shows the particle velocity amplitude patterns of the standing waves for the lowest resonances in an ideal case. In principle a conical tube open at the wide end and closed at the narrow acts as a half-wave resonator, so the resonance frequencies are similar to a harmonic series, in which the second resonance is about one octave above the first. The fact that the resonances are not completely harmonic is because the cone is truncated. An instrument using this kind of resonator is the oboe, which produces the octave when overblown.

Wind instrument resonators mostly deviate from an exactly cylindrical or conical shape. Whether the tube is straight or curved does not matter a lot as long as the curve is not too abrupt. However, the exact cross-sectional inner area is highly significant for the resonance frequencies and hence also for the tuning.

The tube resonator in some wind instruments is partly conical, partly cylindrical, and terminated by a bell. The shape of the bell is of great importance. The

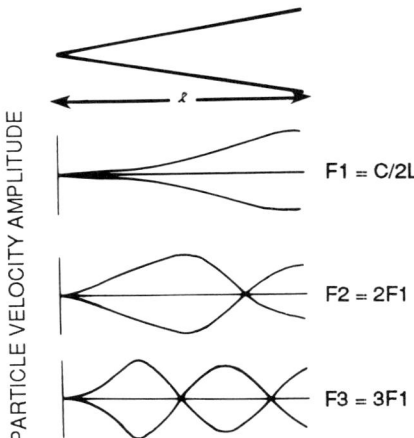

Figure 5.5 Standing waves of the three lowest resonances, represented in terms of the amplitude of the air particle velocity, in a conical tube resonator of length L. For all resonances the velocity amplitude is zero at the narrow, closed end and maximum at the wide, open end.

brass instruments are terminated by a bell belonging to the Bessel horn type, which reflects sound. (The so-called exponential horns have entirely different sound radiation properties; they are used in some types of loudspeakers because rather than reflecting they radiate sound excellently.) Lip-excited wind instruments also have a mouthpiece. It consists of a small kettle-shaped funnel terminated by a short, narrow, and conical piece of a tube.

Bells are very clever at reflecting sound, i.e., they return the sound energy thrown into them from the tube resonator. As a matter of fact, almost all sound energy available in the instrument is used for the purpose of fundamental frequency control. There is only a minor—although well-sounding!—fraction that is leaking out, constituting the sound that we listen to.

The point of reflection in the bell depends on the frequency, and the reflection capability is totally gone above a certain frequency region called the *cutoff frequency*. This frequency is determined by the end diameter of the bell: the narrower this diameter, the higher the cutoff frequency. Another factor of great relevance is how quickly the diameter changes along the length axis. Above the cutoff frequency, the bell cannot reflect any sound so it radiates all sound energy fed into it. In other words, it acts as a megaphone. As a rule of thumb, lower frequencies are reflected far back in the bell and higher frequencies closer to the outer end. Another possibility is to say that the effective length increases with frequency, as illustrated in Figure 5.6. The resonance frequencies are thus affected in a rather complex way when a bell is added to a tube resonator.

In reality the resonance frequencies are influenced not only by the bell and the

II. Tube Resonators

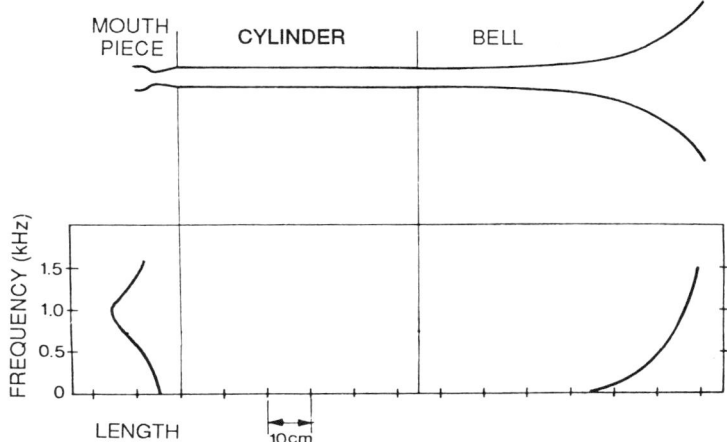

Figure 5.6 Dependence of the effective length on the frequency in a horn. The effective length is influenced in different ways by the mouthpiece and the bell. (From Jansson, 1977.)

bore, but also to a significant extent by the mouthpiece. Also the effective point of reflection in the mouthpiece is frequency-dependent, so that the effective length increases with rising frequency. This is also illustrated in Figure 5.6. By adding a mouthpiece, one changes the resonance frequencies in a rather complex manner.

The net result of these influences on the resonance frequencies is surprising. Tubes provided with both bells and mouthpieces have a complete set of resonance frequencies instead of a set lacking the even-numbered resonances as in quarter-wavelength resonators (2, 4, 6, etc., times the lowest resonance frequency). In addition, the resonance frequencies are rather close to harmonic. The lowest resonance, however, is an exception. It lies way off from its value in a harmonic series. Figure 5.7 illustrates what is happening.

In lip-excited instruments the fundamental frequency is controlled by the resonance frequencies, and the spectrum is harmonic. To allow proper tone production, the resonance frequencies must be reasonably harmonic, as we will see later. This means that it is almost impossible to blow a note with fundamental frequency equal to the lowest resonance on lip-reed instruments; the upper resonance frequencies are too far from being harmonic. However, a tone with the second resonance frequency as fundamental is generally a welcome candidate in the system as are the third, fourth, and some more of the upper resonances. It is thought-provoking that the trumpet has no more than three valves; this means that there are a number of resonances that can be used for controlling fundamental frequency.

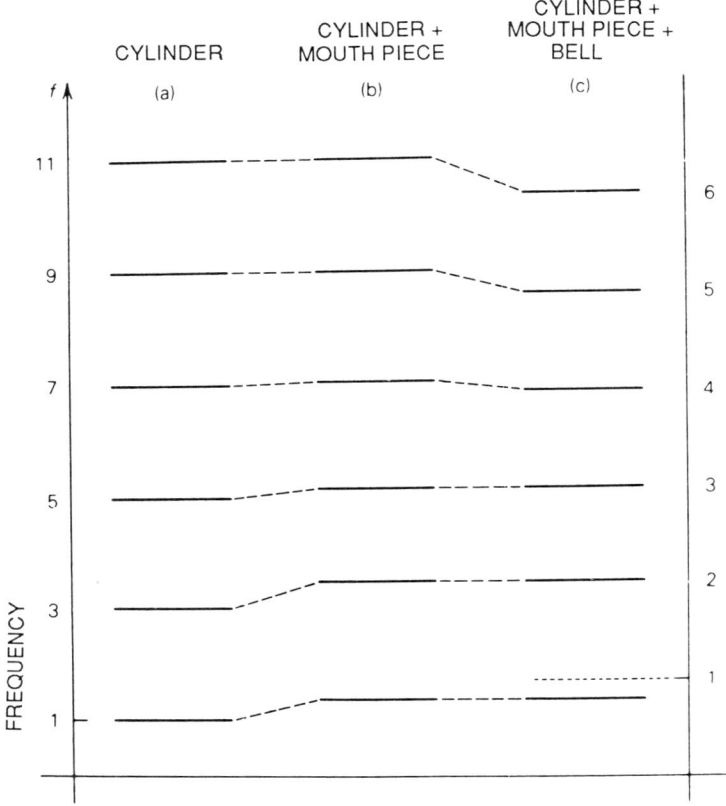

Figure 5.7 Resonance frequencies in a cylindrical tube without extra additions (*left*), provided with a mouthpiece (*middle*), and provided with both a bell and a mouthpiece (*right*). End result is that resonance frequencies become similar to a harmonic series except for the lowest one. (From Hall, 1980.)

III. NO-FEEDBACK INSTRUMENTS

For the no-feedback instruments, there is, as mentioned, practically no feedback from the resonator to the oscillator mechanism. The human voice and the reed stops in the organ, which have short resonators, are examples of no-feedback instruments. Accordion, mouth organ, and harmonium are other examples.

A. Singing Voice

In production of voiced sounds with the human voice, the vocal folds act as the oscillator valve and the pharynx and mouth, sometimes complemented by the nasal cavity, act as the filter.

III. No-Feedback Instruments

1. Voice Source

When voiced sounds are produced, the volume of the air in the lungs is slightly decreased, so that an overpressure of air is produced. In normal speech a pressure of 0.5 kPa is used.

Let us digress here a bit and contemplate the Pa measure. Before we used it for measuring sound pressures, and in Chapter 2 we saw that, roughly, 1 Pa is the pressure generated by an apple over 1 m^2. Traditionally, the driving pressures for wind instruments have been measured in a centimeter water column (cm H$_2$O). This measure goes back to the method of measurement. One bends a tube in a U-shape and fills it half full with water. Then, one applies the pressure to one tube end so that one water column is pressed down and the opposite is pressed up. The pressure is measured as the level difference between these columns. One kPa equals 10 cm H$_2$O.

The overpressure generates an airstream through the slit between the vocal folds. This slit is called the *glottis*. When air is pressed into such a slit with very mobile edges, these edges are brought to vibration. The reason can be described in the following, somewhat simplified way. The airstream throws the vocal folds apart, and their elasticity strives to bring them back. Also the airstream generates a sucking force along the vocal folds, which helps them move toward the midline and eventually close the glottis. When the glottis closes, the airstream is, of course, arrested, but then the elasticity of the folds cannot keep them together, and there is no sucking force from the airstream. Therefore, the glottis opens again, and the process is repeated. In this way, the airstream from the lungs brings the vocal folds to vibration. When we blow straws of grass so that they squeak, we use the same effect.

The resulting regular sequence of air pulses produces vigorous and regular air pressure variations above the glottis. In other words, sound is generated, and this sound is called the *voice source*. Its spectrum is harmonic, and normally it has many partials. It can be varied in several different ways: frequency, amplitude, and the relative strength of the fundamental.

The fundamental frequency depends on the length and vibrating mass of the vocal folds, and these properties are controlled by the musculature in the larynx. The longer and thinner the vocal folds, the higher the pitch.* In normal speech the fundamental frequency for adult men is in the vicinity of 110 Hz and a bit less than an octave higher for adult women. Roughly, a bass singer is able to sing at 65 Hz (pitch C2) and up to 330 Hz (E4). A tenor is working in the region of about

---

*Note that this is a bit alien to what is normally found in music instruments in which "long" is associated with low tones. This depends on the fact that the vibration frequency of the vocal folds is controlled by their mass and tension. The higher the tension, the higher the vibration frequency. If the tension is increased, the folds get thinner, and this adds to the same effect. At the same time the normal relation between size and frequency exists between individuals. Thus, adult males have longer vocal folds than adult females, and bass singers have longer vocal folds than tenors.

123–520 Hz, an alto between approximately 175 and 700 Hz, and a soprano between about 260 Hz and 1,300 Hz.

Fundamental frequency is not entirely dependent on the vocal fold adjustment. Also the air pressure in the lungs is relevant: For each tenth of a kPa pressure increase, the frequency rises by a few Hertz. This dependence on the air pressure in the lungs explains why second-rate singers cannot sing high notes softly: Their vocal fold muscles are not forceful enough to produce the tension needed. Instead, he or she uses the air pressure. However, an increased air pressure does not only result in a higher fundamental frequency, but it also produces an increase in loudness. This effect of an air pressure increase on the voice pitch also explains why speakers mostly raise pitch when they raise vocal loudness; the higher loudness requires a higher pressure, and increased pressure raises the pitch.

Vocal loudness is controlled from the voice source by means of the air pressure in the lungs, as is the case also with other wind instruments. In normal speech, this pressure is typically 0.5 kPa, as mentioned, and in loud speech it is more like 1 or 1.5 kPa. Singers singing high notes in fortissimo may use 3, 6, or even 10 kPa. The acoustically relevant effect of these pressure variations is that the termination of the glottal air pulses is affected. For high pressures, the termination is steep, as is illustrated in Figure 5.8.

The voice source can be varied not only with respect to the fundamental frequency and vocal loudness. Also the relative amplitude of the fundamental can be varied. This results from a variation of the force by which the vocal folds are pressed together. If they are forcefully closing the glottis, a high pressure is needed, and the voice sounds *pressed,* or tense, strangled. In extreme cases, the voice sounds as when one speaks while lifting a very heavy burden. In the other extreme, when the glottal closure is very loose, the resulting voice is *breathy,* as in a voiced whisper. On the way from normal to voiced whisper, there is a voice type that has been called *flow phonation,* because it demands a particularly generous airflow. The vocal quality of flow phonation is characterized by a high amplitude of the voice source fundamental and a great number of higher overtones. Flow phonation was typically practiced in the old days on theater stages and can also often be heard when first-rate opera singers sing, and sometimes also when they speak.

The human voice source consists of a spectrum containing quite a number of overtones. At normal vocal loudness, the spectrum envelope typically falls off at a rate of about 12 dB/octave as measured in flow units. Thus, the second partial is about 12 dB weaker than the first and the fourth is 24 dB weaker than the first. A common characteristic of all wind instruments is that an increase of the blowing pressure not only raises the loudness but also reduces the steepness of the spectrum envelope slope. In other words, the higher overtones gain more amplitude than the lower ones when loudness is increased.

The voice source of a singer differs from that of a nonsinger in many ways. The

III. No-Feedback Instruments

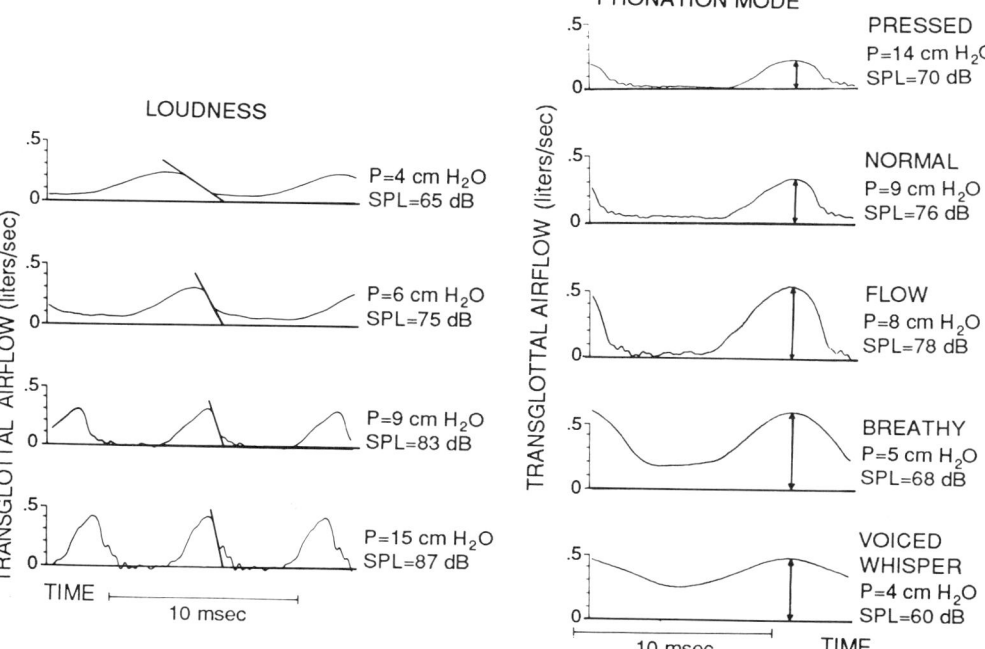

Figure 5.8 Voice source at different degrees of vocal loudness, i.e., at different lung pressures (*left*), and for different modes of voice use generated by varying the pressing together of the vocal folds (*right*). Pressures (P) and resulting SPL are given to the right of each waveform.

most important difference regards how the voice source changes when the pitch and/or loudness is shifted. In a normal speaker, soft tones generally have fewer and weaker overtones, whereas the voice source of a singer is more wealthy in overtones and thus more like the louder tones. Under conditions of high pitch and loud voice, nonsingers' voices generally get tense or pressed because they habitually press the vocal folds together. This reduces the relative amplitude of the fundamental. Good opera singers seem to avoid this. Another important voice difference between a nonsinger and a singer is that many voice source characteristics tend to vary more or less automatically in the nonsinger's voice when pitch or loudness is changed, whereas singers are free to model their voice sources as they please under the same conditions.

All the voice source partials propagate through the pharynx and mouth, or the *vocal tract,* which acts as a resonator. Thus it filters the spectrum of the voice source before it is radiated from the lip opening. The vocal tract resonances, called *formants,* have a great influence on voice sounds: Those voice source partials that are closest to these formants become louder than other partials. The vocal tract resonances or formants therefore appear as peaks in the spectrum envelope (cf. Figure 5.9).

As the vocal tract is practically closed most of the time at the glottal end and open at the lip end, it acts as a quarter-wave resonator. In male adults its length is about 17 cm. As the speed of sound at a body temperature is about 350 m/sec., 17.5-cm vocal tract length is a more appetizing value in calculations. If the vocal tract had been cylindrical, the resonance or formant frequencies would appear near

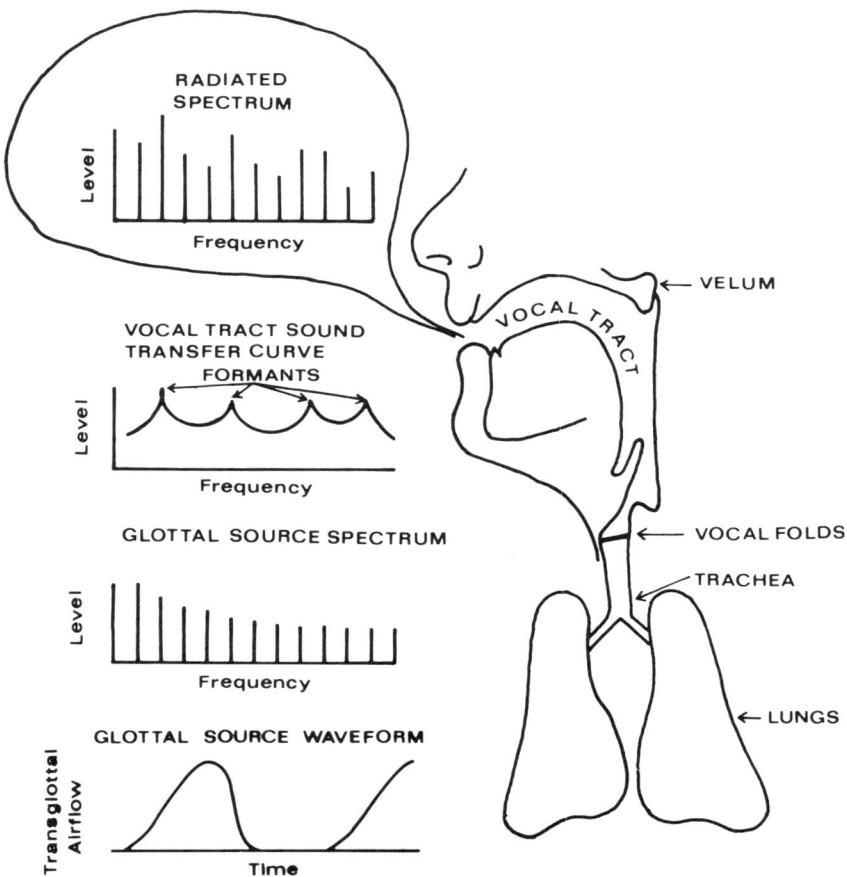

Figure 5.9 Principle for generation of voiced sounds in the human voice. Vocal fold vibrations convert the steady airstream to a pulsating airstream, which implies that a complex tone is produced with an evenly falling spectrum envelope. This spectrum is filtered by the frequency curve of the vocal tract. Hence partials closest to vocal tract resonance frequencies dominate the radiated spectrum. (From Sundberg, 1986.)

III. No-Feedback Instruments

$$f_n = \frac{(2n-1) \cdot 35{,}000}{4 \cdot 17.5} = (2n-1)500 \qquad (n = 1, 2, 3, 4, \ldots)$$

i.e., 500, 1,500, 2,500, Hz [see Eq. (5.3)].

In reality, the vocal tract is far from cylindrical. The consequence is that the formant frequencies depart considerably from the above values. Mostly, the lowest formant frequency, determined by the vocal tract shape, is higher than the voice fundamental frequency that is determined by the vocal fold vibration frequency, as mentioned; we recall that the vocal fold vibration frequency is practically independent of the vocal tract resonator.

It was mentioned that vocal tract resonances are called *formants*.* These have a key function in the production and perception of human speech and song. To cement the formant concept firmly into our arsenal of familiar terms and also to explore their relevance, let us perform a simple experiment. Whisper the following nonsense verse with long, sustained vowels:

    heard - heed - had -heard
    heard - had - heed - heard
    heed - had - head -heard
    heard - had - heed - heard

We may recognize something similar to the Big Ben melody:

How could that melody occur? The reason is the following: When we whisper, we replace the pulsating airstream through the glottis by a turbulent airstream that generates noise. Noise is a sound containing all frequencies, but the amplitude of the noise radiated from the lips is strongest at the formant frequencies. During whispering, the second formant is most clearly perceived; it appears as a somewhat blurred pitch. During whispering there is no equivalent to the pitch corresponding to the fundamental frequency of vocal fold vibration. Rather, the pitch perceived from whispering corresponds to the second formant frequency. By choosing an appropriate sequence of vowels when whispering, one can imitate melodies. In our example, the melody was approximated by the second formant frequency of the vowels.

We may say that the voice source in whisper is the noise produced by a turbulent

---

*This is the most appropriate way of defining the term *formant*. As a consequence, formants can occur, strictly speaking, only in sounds generated by the human voice organ. However, in some books and articles the reader may meet formant definitions associated with spectrum envelope peaks, which are mostly unaffected by a change of fundamental frequency.

airstream. The turbulence occurs when the airstream passes the narrow slit between the vocal folds, which in this case are so stiff that they refuse to vibrate. The airstream is made turbulent by passing such narrow slits also in voiceless consonants such as t, s, and p. Thus, in all these sounds, the source is noise.

2. Formants and Articulation

Earlier it was mentioned that the resonance frequencies in a tube resonator depend on its length and shape. In the vocal tract these parameters are controlled by *articulation*. By squeezing the tube in one place and expanding it in another, we can make the resonance frequencies deviate substantially from the harmonic series 500, 1,500, 2,500, ... Hz that we talked about earlier. By means of articulation an adult male can vary the frequency of the first formant between approximately 150 and 900 Hz, the second between 500 and 3,000 Hz, and the third between 1,500 and 4,500 Hz.

When we whispered the nonsense verse, we varied the lip and jaw openings and the tongue shape from vowel to vowel. These are the means used for varying formant frequencies. As rules of thumb for the relation between articulation and formant frequencies, the following applies.

- A narrowing of the lip opening and an increase of the vocal tract length, produced by lip protrusion or larynx lowering or both, lowers all formant frequencies more or less.
- A widening of the jaw opening tends to increase the first formant frequency.
- The shape of the tongue affects the frequency of the second formant in particular. If the tongue body is pushed upward-forward, narrowing the front end of the vocal tract, the second formant frequency is raised, and if the middle part of the vocal tract and the lip opening are both narrowed, the effect on the second formant is the opposite.
- The tongue tip is particularly efficient in changing the third formant frequency. More specifically it is the small cavity between the lower incisors and the tongue tip that greatly affects this formant in most vowels.

Because the partial closest to a formant becomes louder than other partials (cf. Fig. 5.9), the spectrum envelope is continually changed as we change articulation. You can convince yourself about this in the following way. Sustain a soft note at the pitch of approximately C4. While doing this, vary the articulation rhythmically between the vowels in who'd, heard, head, heed, head, heard, who'd, etc. By continuing this exercise for a while, you will soon notice an emerging triad melody. (When writing this, I practiced a little, with the result that my little son came running, asking what the funny sound was.) The effect results from the fact that the second formant visits different adjacent partials in the harmonic spectrum. Partials number 4, 5, and 6 constitute a major triad, as we recall from Chapter 4.

Note that there are two distinct aspects of timbre of vocal sounds. One is the

III. No-Feedback Instruments 123

vowel quality, which determines which *vowel* we perceive, e.g., if it is an ee or an aa etc. Another aspect is the *voice quality,* which determines, for example, whether it sounds as if Mr. Jones or Mr. Smith pronounces the vowel. Thus, the voice quality is a personal characteristic. These two aspects of the timbre of vowels should not be confused.

For most vowels the two lowest formants determine the vowel quality. Therefore the timbre quality that is characteristic for a vowel can efficiently be described in terms of the frequencies of the two lowest formants. Figure 5.10 offers an example of this kind of description. It shows the frequency ranges for these formants in different vowels in the form of a diagram, where the first formant frequency is plotted on the horizontal axis and the second formant frequency on the vertical axis. We can see that the vowels in the words heed, who'd, and hud are extreme and constitute the corners in a somewhat triangular contour. The

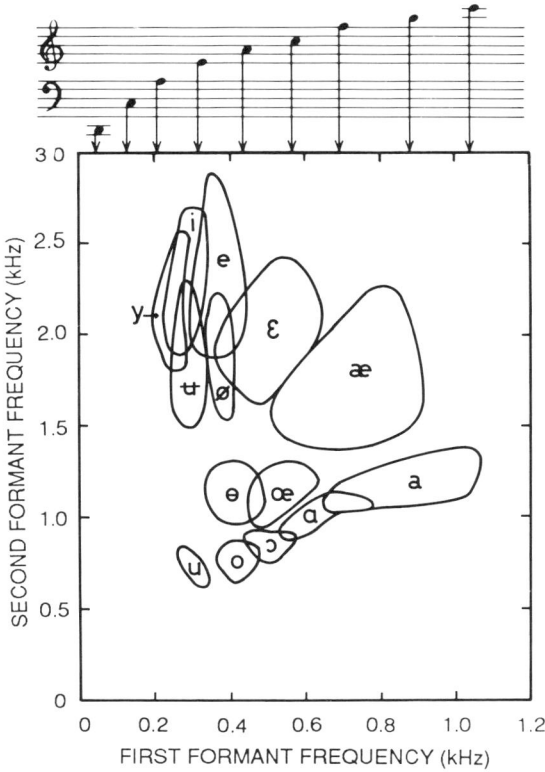

Figure 5.10 Two lowest formant frequencies of Swedish vowels. Symbols: u, booed; o, bode; ɔ, box; ɑ, pod; a, German Tag; æ, bad; ɛ, bed; e, bayed; i, bead; y, French tu; ʉ, Swedish bud; ø, German schön; ɵ, Swedish ull; œ, German zwölf.

second formant frequency is highest in heed and lowest in who'd, whereas the first formant frequency is highest in hud and lowest in who'd and heed. As mentioned, you can hear the relative frequency location of the second formant in terms of a somewhat diffuse pitch by whispering the vowels.

In the same figure we can also see that the distances between adjacent vowels in the diagram vary considerably. The vowels in heed and head are quite clustered, whereas those in who'd and hawk are well-separated. It is interesting that these intervowel distances would be more even if tone height rather than frequency were used in the diagram, i.e., if mel rather than Hertz were used as the unit.

The second formant frequency appears to have a special significance to our perception of timbre. If we hear a sound with dominating low-frequency components, as, for example, the sound of a bass tuba, we often would call the sound "too-too." If the dominating components are a bit higher in frequency and loud, the sound would rather be called a "bang," and if they are really high, "beep" would seem more appropriate.

The formant frequencies are influenced by the vocal tract length: the longer the tract, the lower the formant frequencies. Adult men tend to have longer vocal tracts than women, and this causes an important difference in the formant frequencies between them. The formant frequency values plotted in Figure 5.10 are true for adults. Children obviously have shorter vocal tracts, and this accounts for a good deal of the timbre difference between ladies' choirs and children's choirs. However, it is mainly the higher formant frequencies, number 3, 4, and 5, that produce the typical voice timbre differences between males, females, and children.

The vocal tract length and hence also the formant frequencies of a given vowel varies somewhat also within the groups of men, women, and children. Such differences explain a good deal of the voice timbre variations among individuals. For example, it has been shown that for a given vowel, tenors tend to possess higher formant frequencies than bass singers. Relatively high formant frequencies are therefore typical for the tenor voice timbre.

Up to now we have mainly talked about vowels, which, of course, are particularly important in singing simply because they tend to be much longer than in speech. The consonants are also formed by formant frequencies but not exactly in the same way as vowels. In consonants, the characteristic aspect is *change* in formant frequencies. Thus, in most consonants the lowest three formant frequencies make typical patterns that we recognize and identify.

3. Singer's Formant

A characteristic feature of bass, tenor, and alto singers' voices is an unusually high spectrum envelope peak occurring somewhere between approximately 2 and 3 kHz. It appears in all voiced sounds. It seems to depend on a clustering of the third, fourth, and fifth formant frequencies. This spectrum envelope peak has been

III. No-Feedback Instruments

called the *singer's formant*. It is illustrated in Figure 5.11. The articulation producing this formant frequency cluster appears to be a widened pharynx, which can be achieved by a lowering of the larynx.

The singer's formant contributes to making the voice timbre more distinct and "shiny." Note also that it appears in a frequency range in which the ear is particularly sensitive (cf. Figure 3.11). The overtones of an accompanying orchestral sound are also much softer in this frequency range than near 500 Hz, where they are generally loudest. The net result is that the singer's formant contributes to making the voice of the singer more easy to hear even when the orchestral accompaniment is loud. It is quite fortunate that this effect can be achieved without any need for excessive recruiting of vocal effort.

4. Formants at Super Pitches

Sopranos have been found to use a different strategy than other singers to produce audible sounds in the presence of a loud orchestral accompaniment. We have probably all noted that they tend to widen their jaw opening with rising pitch, almost regardless of what the vowel is. This seems to hold at least up to frequencies near A4 (880 Hz). In this manner they raise the first formant frequency in step with the fundamental frequency. The adaptation of the vocal tract shape to the fundamental frequency affects all formant frequencies as shown in Figure 5.12. At the top pitch, all vowels share approximately the same formant frequencies for the soprano represented in that figure.

Probably, the reason for these formant frequency changes is acoustic. The first formant frequency is tuned to a frequency slightly above that of the fundamental as soon as, otherwise, it would be lower than the fundamental. As a result, the amplitude of the fundamental is increased for reasons of resonance, and the sound

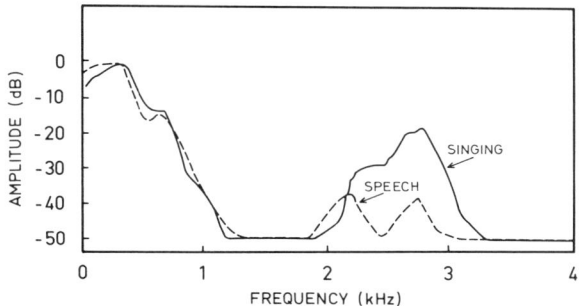

Figure 5.11 Spectrum envelopes of the vowel in the word *booed* as pronounced by a male baritone singer as in normal speech (*dashed curve*) and in singing (*solid curve*). (From Sundberg, 1986.)

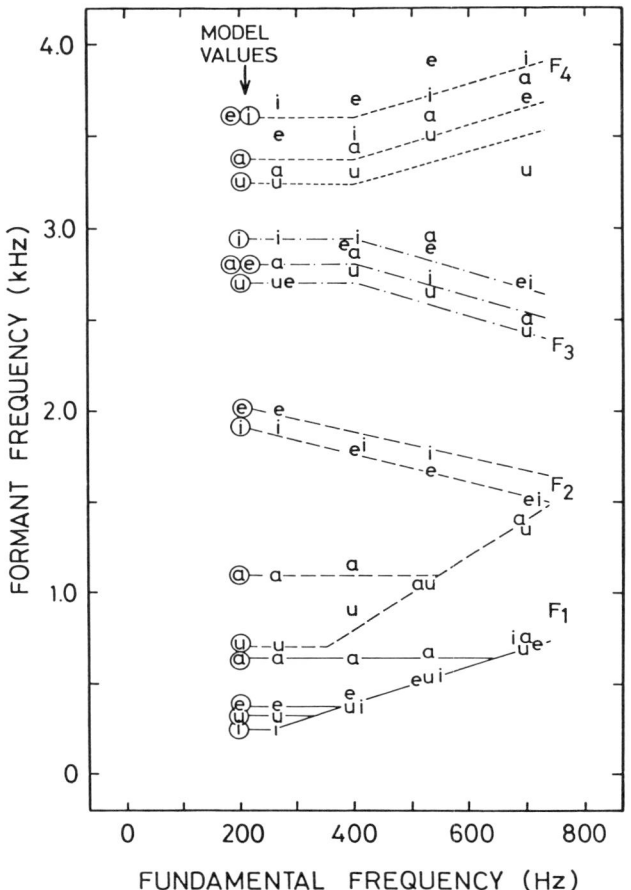

Figure 5.12 Formant frequencies for various vowels as produced at different fundamental frequencies by a professional soprano. Circled and uncircled vowel symbols show measurements for spoken and sung and the lines represent an idealization. (From Sundberg, 1986.)

level of the radiated sound is accordingly raised. When the singer sings at high pitches, the sound level she produces increases considerably if she increases her jaw opening to an appropriate extent. Probably not only sopranos are applying this strategy. Depending on the normal frequency of the first formant of the vowel and the pitch sung, the strategy is indicated also for many vowels sung at the upper alto pitches and those sung at high tenor pitches.

We mentioned that the vowel quality is determined mainly by the two lowest formant frequencies. Therefore, one would expect disastrous consequences regarding vowel intelligibility in the cases mentioned. Facing the choice between inaudible tones with normal vowel quality or audible tones with strange vowel quality, singers probably pick the right alternative. Also, the vowel quality of sustained vowels survives these pitch-dependent formant frequency revisions surprisingly well, except for the very high pitches, at fundamental frequencies above about 700 Hz. Above that frequency, no formant frequency combination seems to help, and below it the vowel quality would not be better even if normal formant frequencies were chosen. The amount of text intelligibility that occurs at these very high pitches relies almost exclusively on the consonants preceding and following the vowel.

5. Vibrato

A characteristic of the trained singing voice is the frequency vibrato, previously mentioned in Chapter 3. In singing, the vibrato corresponds to a slow, nearly sinusoidal variation or modulation of the fundamental frequency. Normally, the rate is somewhere between 5 and 7 undulations per second, and the modulation depth varies between +/−50 and +/−150 cents. Both the rate and the depth are important. If the rate goes below five undulations per second, which may occur in old and/or strained voices, no clear single pitch can be perceived; it sounds as if the pitch is swaying around. If it is faster than seven undulations per second, the tone sounds nervous; singers sometimes call this a *rabbit vibrato*. Also if the modulation depth surpasses decorum, the vibrato sounds exaggerated.

The partials close to a formant frequency vary in amplitude depending on its distance to that formant. As a consequence, an amplitude vibrato is generated. This byproduct is of minor perceptual significance.

The origin of the vibrato is not known. It has been demonstrated that the vibrato corresponds to pulsations in the neural control signals going to various laryngeal muscles, those associated both with fundamental frequency control and with closing the glottis. It has also been revealed that the airflow tends to vary in synchrony with the vibrato, which is a natural consequence of the pulsations in the muscles regulating how forcefully the vocal folds are pressing against each other. What causes these pulsations in the neural signals is uncertain. Some assume that it is similar to tremor in the laryngeal musculature; tremor is a similar pulsating contraction of muscles, often appearing after some time of constant contraction.

B. Other No-Feedback Instruments

The common feature for the singing voice and other no-feedback instruments is, as was mentioned before, that the fundamental frequency is controlled not by the resonator but within the oscillating valve mechanism. Thus, the pitches are tuned by adjustments affecting the mechanical properties of the source. In these instruments the oscillator valve is a metal blade applied so that it can alternately allow and arrest an airflow into the resonator, either by swinging out and into an opening to the resonator, as in free-reed instruments, or by beating against the opening, as in most of the reed pipes in the organ (beating reed). The metal reed is brought to vibration by the airstream, so that it periodically cuts the airstream, and a periodically pulsating airflow is generated.

In these instruments the vibration frequency is almost entirely determined by the mechanical properties of the oscillator valve. The reed pipes in the organ are tuned by adjusting the vibrating length of the reed: the longer the reed, the lower the tone (note the difference with the human vocal folds!) In some bass reeds, a little lead mass is fastened to the free end of the reed. This increase of the reed mass lowers the frequency, so that all the frequency lowering need not be taken care of by the reed length. This is, of course, advantageous in instruments that sell well if they have small dimensions. In the harmonium, mouth organ, and accordion, the length of the tongue is adjusted once and for all in the factory, and this tuning then is supposed to last as long as the rest of the instrument.

In the regal and in its organ stop cousins, the resonator is a short piece of tube. The lowest resonance frequency is *higher* than the fundamental frequency, as is normally the case with the human voice. The resonance frequencies are determined by the resonator shape, and they impose peaks in the spectrum envelope, as in the human voice.

Because the construction and function of the regal pipes are basically the same as in the voice organ, it is often easy to mimick their timbre using the voice. There are good acoustic reasons behind the fact that the stop *Vox Humana* (Latin for the human voice) is a regal stop.

Many no-feedback instruments such as harmonium, mouth organ, and accordion lack proper resonators. Instead, some of them have small and open boxes as resonators. These resonators are heavily damped, and thus do not shape the radiated spectrum to any great extent. They seem to radiate the source spectrum rather faithfully.

IV. FEEDBACK INSTRUMENTS

A. Input Impedance

In feedback instruments both the fundamental frequency and a good deal of the spectrum are dependent on the properties of the resonator. A great enthusiast and authority within music acoustics devoted an almost lifelong and very productive research activity on wind instrument acoustics. His name was Arthur H. Benade, professor of physics at Case Western Reserve University in Cleveland, Ohio, and also an accomplished clarinet player. His theory can be summarized as follows.

IV. Feedback Instruments

An important aspect of the resonance properties of the wind instrument resonator is its highly frequency-dependent willingness to acccept input sounds. As a matter of fact, the sound appetite at the end of a tube resonator varies dramatically because of resonance. This sound appetite, or, more specifically, the willingness to accept vibration at different frequencies, is called the *input impedance*. It is defined as the ratio between the resulting sound pressure amplitude and the input volume velocity amplitude which is a measure closely related to the particle velocity mentioned before; it specifies what volume of air is transported through an area per time unit, so the unit is cubic meters per second rather than meters per second.

The input impedance is measured approximately in the same way as resonance frequencies and bandwidths. The difference is that for input impedance measurements, the sound source and the microphone are placed in the same end rather than in opposite ends.

What an input impedance curve shows is how high the sound pressure amplitude gets for a given input airflow amplitude. At frequencies in which the input impedance is high, the willingness of the resonator to oscillate is high at that end, because already at small amplitudes of the input sound the resonator responds with a great pressure amplitude. A moment's reflection will convince us that this input impedance must be highly significant for the sound generation in a feedback instrument. In such instruments, the input is an oscillating airflow from an oscillating valve, and the goal is great pressure oscillations. The question is at which frequencies this is going to happen.

Figure 5.13 shows some examples of input impedance curves for different types of resonators. We can see that the curve is a bit similar to the sound transfer curve

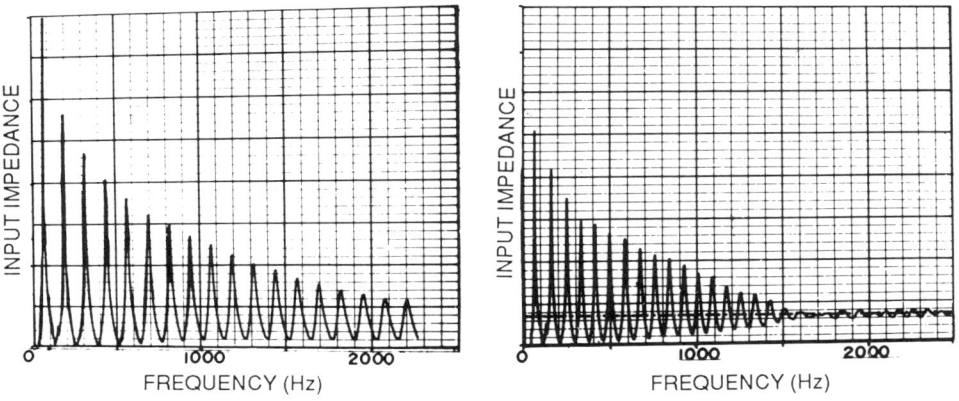

Figure 5.13 Input impedance curves for two different tube resonators. The *left panel* refers to an open-closed cylinder, the *right panel* a cylinder provided with a bell. (From Benade, 1976.)

shown in Figure 5.1, although the frequencies were higher in that case. The factor deciding which tones can be blown and which partials can be generated is how high the input impedance curve reaches *at the frequency of each single partial*. One should keep in mind that the input impedance value at each single frequency actually shows how much the resonator likes to oscillate at that particular frequency. In case the resonator hates vibrating at some frequency, it will not be easy to convince it to do so.

A highly important circumstance is the fact that the partials are *exactly harmonic* in wind instrument spectra; they occur as a consequence of the fact that a sine curve is knocked out of shape. A wind instrument always tries to oscillate at a frequency such that a maximum number of all the harmonic partials should feel reasonably welcome in the resonator. This implies that the instrument will favor a fundamental frequency such that the input impedance is sufficiently high at the integer multiples of that fundamental frequency. This is the very key to the sound generation in wind instruments. In these instruments many partials are enjoying a reasonably high input impedance.

B. Fundamental Frequency Control

The control of fundamental frequency takes place in the following manner in wind instruments. The player forces an airflow through the slit between the lips, between the reeds, or between the reed and the resonator edge. Then the slit is closed for approximately the same reasons as in the case of the human vocal folds. A pulse of overpressure is produced, which propagates through the resonator tube. At its open end, most of the pulse is reflected and returns toward the oscillator. When it arrives there, it changes the pressure so much that the slit opens. This generates a new pulse, and the process is repeated.

The travel time that the pulse needs for the trip oscillator–reflection place–oscillator depends on the speed of sound propagation c and the effective tube length l_e. As travel time is distance divided by velocity, the following applies for a half-wave resonator:

$$T = \frac{2l_e}{c} \tag{5.4}$$

Because $T = 1/f$, the vibration frequency of the lips can be computed:

$$\frac{1}{f} = \frac{2l_e}{c} \quad \text{or} \quad f = \frac{c}{2l_e} \tag{5.5}$$

which is the fundamental frequency of the lowest possible tone. If the resonator is of the quarter-wave type, the lowest possible tone is one octave lower, because other resonance properties apply.

IV. Feedback Instruments

The resonator does not alone determine the fundamental frequency. Also the oscillator valve has limitations as to its willingness to vibrate at different frequencies, i.e., it has a resonance frequency of its own. The player adjusts this resonance by varying his or her lip tension, depending on the tone intended.

The speed of sound propagation in Eqs. (5.4) and (5.5) depends on the temperature of the resonating air column (see Chapter 2.III). This is in turn dependent on the temperature of the player's expiratory air and the temperature of the resonator walls. For this reason, a feverish wind instrument player would tend to play a little sharp as compared with a perfectly healthy one. The temperature of the resonator walls are heated during the playing to a stable value, but as long as the instrument is cold, the sound propagation is slower and the tuning a bit flatter than otherwise. Recorder players, who frequently switch between various instruments while playing, therefore often excel by fishing up one flute while burying another one in sleeves, sweaters, and shirts, the entire maneuver being completed during an eighth note's rest or so. Thus, it is essential to store the instruments at body temperature when there is no time to warm up the instrument before starting to play it.

What are the locations of the input impedance peaks of tube resonators? Their exact frequency locations equal the resonance frequencies of the system and are thus determined by the shape of the tube. When it comes to the tuning of instruments, the last Hertz may be essential. This implies that the instrument shape is significant in its finest details. Minute pieces of dirt may be sufficient to have an effect on a tone.

To vary fundamental frequency in feedback instruments, it is apparently necessary to change the resonance frequencies of the resonator while playing. This is achieved by varying the resonator length or, strictly speaking, the effective length of the resonator.

The main principle is to change the tube length by lengthening or shortening the instrument. Lengthening can be achieved by opening a valve that causes the air to pass through an extra piece of tubing so that the overall length of the duct is changed. There are generally several extra pieces of tubing so that a number of different lengths are available. In the trombone a narrower tube is slid in and out of a slightly wider tube, so that the length can be varied in this manner. In some instruments, such as the old curved wooden trumpet, called the cornett (German Krummer Zink) and serpent, there are side holes instead, as in the flute.

Shortening of resonators can also be achieved by opening side holes. In principle, the effective length of the resonator ends slightly beyond the first open hole. However, if this hole is narrow, the extra length may be substantial. In some instruments, there is a hole just at the location of a pressure maximum of the longest standing wave. This hole is operated by a key, the octave key. By so doing, the possibilities for this standing wave to develop are eliminated, so that only other standing waves corresponding to higher resonance frequencies remain. This method of tuning the resonator is called *overblowing,* a term that was used several times in the beginning of this chapter; it is used in cane and air reed instruments.

If the resonator is of the half-wavelength type, overblowing will give the octave, because in this case the fundamental frequency is controlled by a resonance approximately one octave above the lowest one. If, however, the resonator is of the quarter-wavelength type, overblowing will give the duodecime (i.e., octave plus fifth), because in this case the resonance controlling the fundamental frequency is about a duodecime above the lowest one.

One way of imagining why overblowing occurs on the octave is that the player injects two air pulses twice as close in time before he or she receives the first one back; as we know, the frequency of the second resonance is twice that of the first in this case. Overblowing on the duodecime and other resonance frequencies can be viewed in a corresponding way. What tone the player hits is determined also by the mechanical properties of the lips, i.e., *their* resonance frequency.

In instruments played with side holes, the point of reflection is in the vicinity of the first open hole, as mentioned. At the same time we know that such instruments have bells. What is the advantage of the bell, when a side hole is open? The answer is, in fact, NIL! One can easily remove the bell, without affecting the tone the slightest little bit if a side hole sufficiently far from the bell is open.

Particularly in lip reed instruments, another type of overblowing is used. It implies that a higher resonance is used for the control of the lip vibration frequency. If the effective length has been set to a certain frequency, e.g., by pressing a key, the resonator is tuned to a particular series of resonance frequencies. By varying the manner of blowing by adjusting the resonance frequency of the lips, the player can select a frequency such that a maximum number of resonance frequencies is met by a sufficiently high input impedance in the resonator. In such instruments, overblowing implies that the fundamental frequency of the spectrum is placed on a frequency other than the lowest resonance frequency.

This type of overblowing is required, when the lowest resonance is way off from its value in a harmonic series, as in the case of lip reed instruments. Then, one must neglect the lowest resonance and address a higher one instead, which can serve as part of a reasonably harmonic series of resonances. If one chooses to place the fundamental frequency on a higher resonance, the auspices look much brighter, as shown in Figure 5.7. In the lip reed instruments, the resonance frequencies above the lowest one constitute a reasonably harmonic series. If one selects one of the lower resonance frequencies above the lowest one as fundamental frequency, a sufficiently high number of partials may face a rather high input impedance.

C. Spectrum Shaping

In a feedback instrument the spectrum is determined approximately in the same way as the fundamental frequency: It is the input impedance that is determinant. If it is low at the frequency of a particular partial, that partial will be radiated with a low amplitude. Another significant factor is, of course, how much the bell or the radiating side holes are leaking at that frequency. If the bell does not want to let a partial out, this partial will remain in the resonator so that it becomes much stronger within the instrument than in the radiated spectrum.

There are minute microphones available that can be inserted into a resonator without appreciably disturbing its properties. By means of such a device one can record the sound standing within the resonator while a tone is being played on the instrument. This *internal spectrum* is filtered by the sound radiator before it is allowed to pass to the air outside the instrument. Observe the difference between this internal spectrum and the voice source. The internal spectrum is the sound traveling inside the instrument, whereas the voice source is the sound generated by the vocal fold vibrations and fed into the vocal tract. The internal spectrum is used to control fundamental frequency, which in the voice is determined by vocal fold adjustment. The spectrum that wind instruments radiate is the internal spectrum filtered by the sound transfer characteristics of the radiating bell or side holes.

A bell reflects the sound within the resonator reasonably well up to its cut-off frequency, which is determined by the dimensions of the bell, as mentioned before. If the end diameter is wide and if the bell flares slowly, the cutoff frequency is low. The effect of the cutoff frequency on the input impedance curve can be observed in Figure 5.13, in terms of the disappearance at 1.5 kHz of the peaks in the curve.

Regardless of the type of wind instrument, the player can modify the properties of the radiated spectrum in various respects. If the blowing pressure is raised, it is not only the sound level that increases, but also the spectrum slope is reduced. Thus, the higher overtones grow more quickly in amplitude than the lower overtones under these conditions, just as in the human voice. Another influential factor is exactly how the instrument is played, e.g., how forcefully the lips hold the reed.

In the case of the trumpet, an interesting relation has been revealed between the amplitudes of the fundamental and the overtones: the stronger the fundamental, the stronger the overtones (see Figure 5.14). For instance, if the pressure amplitude of the fundamental increases by a factor of 100 (40 dB) that of the second partial increases by a factor of 160 (44 dB). The increase is greater for the higher overtones; for the seventh partial the ratio is 400 (52 dB). During the onset, the spectrum is entirely dominated by the lowest partials, whereas the higher partials have rather low amplitudes. The principle is illustrated schematically in Figure 5.14. If applied in synthesis, the result sounds exactly as a trumpet, particularly if the amplitude of the fundamental grows in a realistic fashion.

D. Cane Reed Excitation

With respect to the source, there are reasons to distinguish between instruments with double and single reeds. The clarinet and the saxophone have single reeds, whereas the oboe and the bassoon have double reeds.

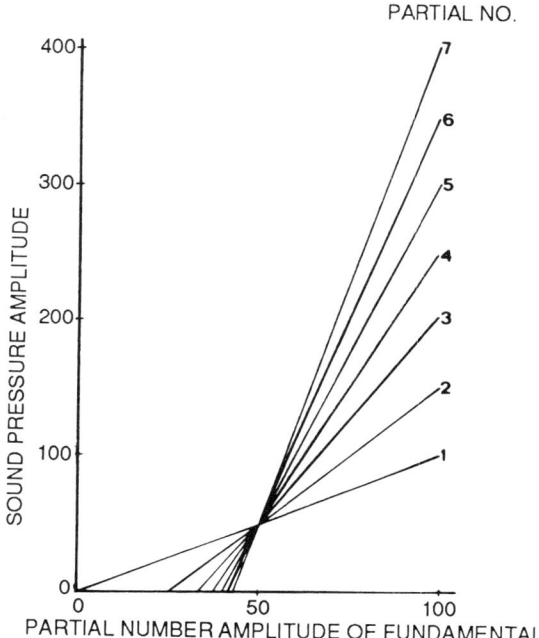

Figure 5.14 Idealized relation between sound pressure amplitudes of the seven lowest partials in a trumpet. A factor of 100 corresponds to 40 dB. Amplitudes of these overtones can be predicted from the amplitude of the fundamental. (From Risset and Mathews, 1969.)

A characteristic feature of the source of instruments with double reeds is that partials lying within certain frequency ranges are particularly strong, irrespective of the frequency of the fundamental. Thus, maxima can be observed in certain frequency regions of the spectrum regardless of which pitch is being played. The bassoon has dominant partials near 500 Hz, and the oboe has dominant partials near 1,000–2,000 Hz, no matter what the fundamental frequency is. Figure 5.15 offers examples.

These maxima in the spectrum envelope originate from the reeds' vibration pattern. For this reason they should not be confused with spectrum envelope peaks of vocal sounds, which derive from the formants, the vocal tract resonances. *Formants* has often been used as a term for these spectrum envelope peaks. However, according to our terminology, formants can only exist in the human vocal tract. Therefore, quasi-formants may be a more appropriate term. It is not very difficult to imitate the sound of a bassoon with the voice. This is because it is not difficult for the voice to tune the two lowest formants so that the same type of broad spectrum envelope peak occurs at approximately 500 Hz.

The clarinet has a single reed, as mentioned, and its spectrum structure is

IV. Feedback Instruments

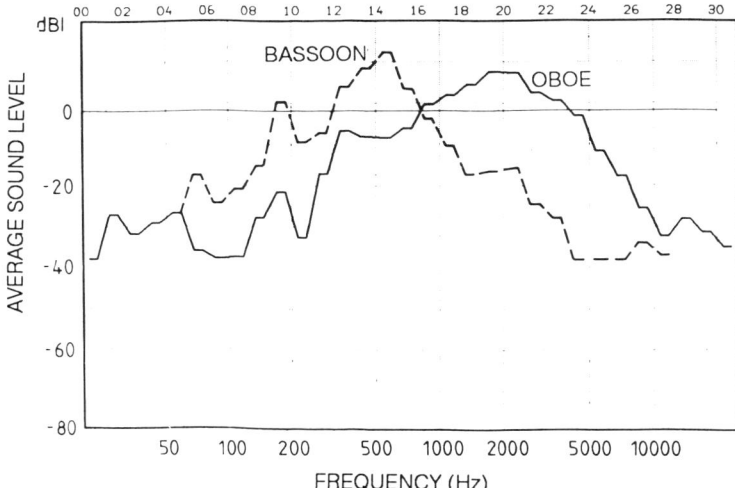

Figure 5.15 Averaged spectrum envelopes for a scale played on an oboe and on a bassoon. The figure shows the average sound level in third-octave frequency bands. The frequency scale is arranged to correspond to the critical bands of hearing.

entirely different from that of the double reed instruments. The lower even-numbered partials, as the second, fourth, and sixth, are substantially weaker than the odd-numbered ones for tones in the lower part of the instrument's frequency range. This has the following explanation.

The resonator of the clarinet is basically a cylindrical open-closed tube. In other words, it has a quarter-wavelength resonator. This means that even-numbered partials are not welcome in the resonator, because the input impedance is low at their frequencies. A common misunderstanding is that these partials are all but missing in the spectrum. The truth is that the second partial may be about 40 dB below the fundamental, so it hardly contributes to the timbre. Higher up in the spectrum these differences between even- and odd-numbered neighbors are smaller. Further, as illustrated in Figure 5.16, the differences can be found only for the instruments' lower tones.

The saxophone, being another instrument with single reed, has a half-wavelength resonator, because the resonator tube is conical. Adjacent partials in the spectrum are therefore of comparable amplitudes.

E. Lip Reed Excitation

This group includes all instruments in which the player's lips provide the excitation. Trumpet, trombone, bass tuba, and cornet are all members of this

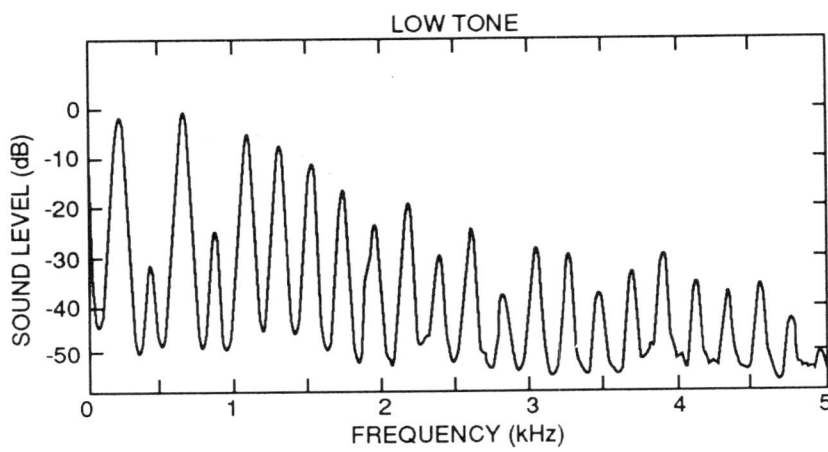

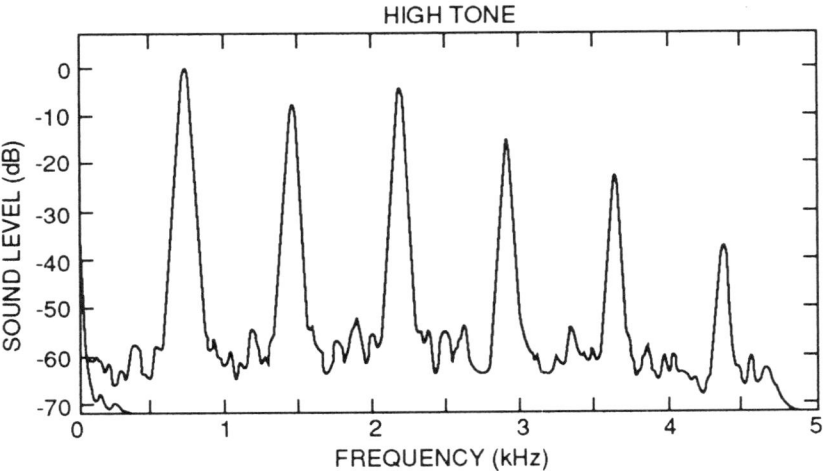

Figure 5.16 Spectrum of a low and a high clarinet tone.

IV. Feedback Instruments

family. These instruments have been mentioned several times above when speaking about resonators and sound generation.

A property of particular significance to the timbre of nearly all lip reed instruments is that all tones are radiated from the bell rather than from side holes, as in cane reed instruments. The bell contributes to shaping the radiated sound because it leaks sound in a frequency-dependent way. This frequency dependence is reasonably simple. It can be specified in terms of the cutoff frequency mentioned above, which, in turn, is determined by the shape and size of the bell.

If the spectrum envelope of the internal spectrum slopes toward higher frequencies, the result is that the radiated spectrum receives a broad, blunt peak near the cutoff frequency of the bell, as illustrated in Figure 5.17. This peak will be higher in frequency if the bell is narrower, because then the cutoff frequency is higher.

For the French horn, the partials near 500 Hz are the most dominating in the spectrum as can be seen in Figure 5.17. These partials are the strongest regardless of the fundamental frequency. The spectrum peak is broader than for a normal formant. However, we can still imitate the sound of a French horn if we tune the first and second formants appropriately and broaden their spectrum peaks by sufficiently increasing the vocal tract attenuation. This can be done by making the lip opening wide and very narrow.

A difference between old and new trombones is that the old bells were smaller than the modern ones. The old baroque trombones therefore produce a spectrum peak centered at a higher frequency. It is easy to tell the difference between these two instrument types, and this illustrates both the significance of the bell for the timbre and the timbral significance of these quasi-formants. Other lip reed instruments largely lack such clear-cut quasi-formants, which appear independent of fundamental frequency.

F. Air Reed Excitation

This group consists of instruments in which an airstream, shaped as a reed, is brought to oscillation. The group includes all flute instruments like flute, recorder, edge flute, and the flue pipes in the organ.

In these instruments the airstream pass through a narrow slit, which imposes the reed shape. This airstream then passes close to an edge, the upper lip in the recorder and the edge of the embouchure in the flute.

The sound generation in these instruments can be described as follows. When this air reed passes the edge of the pipe's mouth, the resonator is excited so that it starts oscillating on its resonance frequencies. This means that the air particles in the mouth (as well as those in the other open end, if any) pass out of and into the resonator with frequencies equal to the resonance frequencies, because, as we

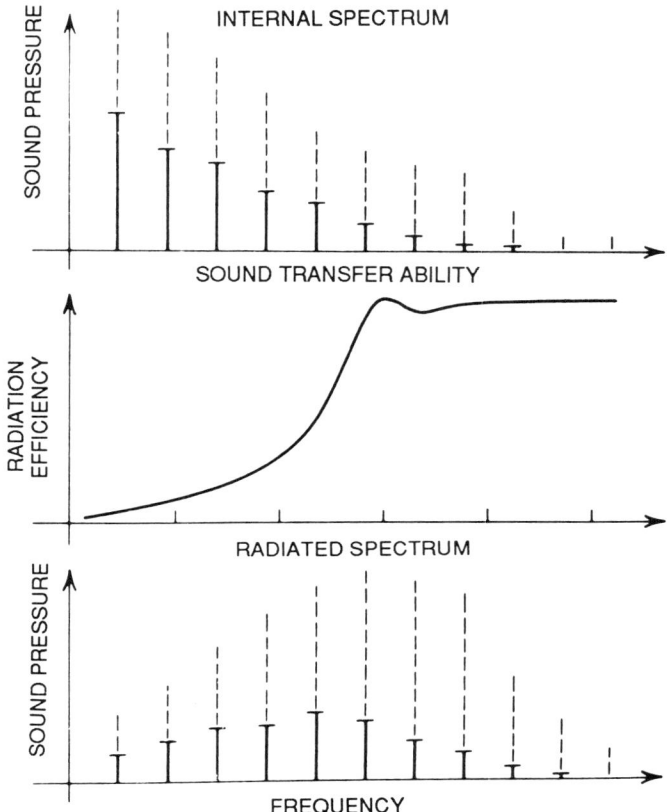

Figure 5.17 Interaction between the internal spectrum traveling back and forth in the resonator between mouthpiece and bell is shown as well as the sound transfer ability of the bell which imposes quasi-formants on the spectrum of lip reed instruments. Solid lines, pp; dashed lines, fff. In the French horn the quasi-formant appears near 500 Hz, regardless of fundamental frequency. (From Hall, 1980.)

recall, sound is equivalent to movement in air particles. The air particles in the mouth then push and pull the air reed and thereby affect its direction; the air particles push the reed into the pipe when they move into the pipe, and vice versa. This increases the excitation. In this way, a resonance takes command over the fundamental frequency and the system generates a spectrum consisting of harmonic partials. Thus, the system is somewhat similar to those of the other feedback instruments.

In most air reed instruments, the resonator is almost perfectly cylindrical, and the resonance frequencies are determined by the shaping of the resonator. As the mouth or embouchure represents a rather wide opening, the resonator is of the

IV. Feedback Instruments   139

open-open type, i.e., half-wavelength resonator. This, in turn, means that there are input impedance peaks near all partials, welcoming them into the pipe.

However, this is not an accurate description of the situation. The shape of an aperture has a great influence on the harmonicity of the resonance frequencies; a narrow aperture in an open flue organ pipe induces a higher harmonicity than a wide one. As such pipes are generally cylindrical, the aperture area equals the cross-sectional area of the tube. In other words, the flue organ pipe diameter is decisive to the harmonicity of its resonance frequencies. More precisely, it is the ratio between the diameter and the length of the pipe that is decisive. Narrow organ

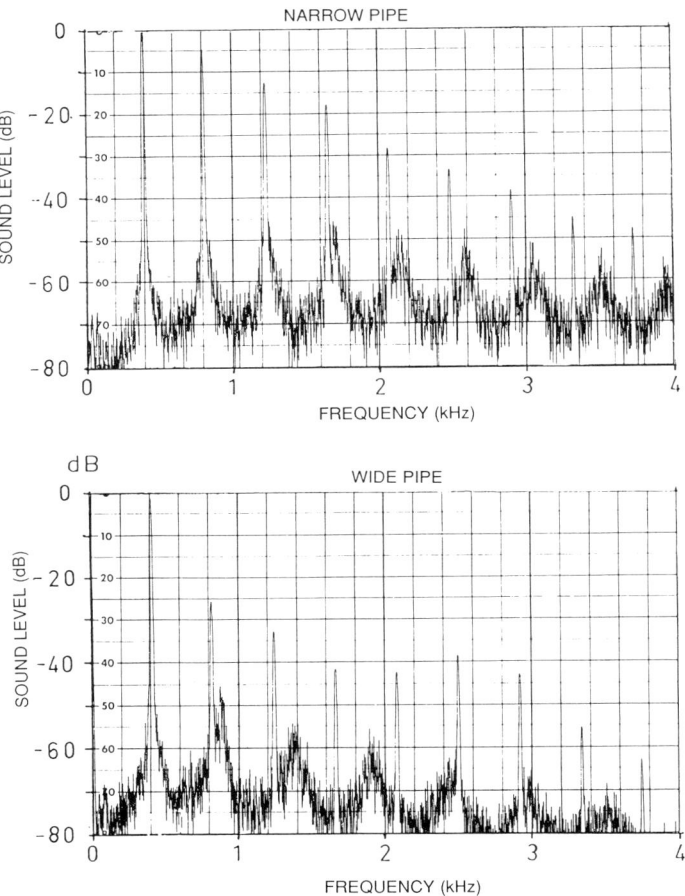

Figure 5.18 Spectra of a narrow and a wide organ pipe. (From Sundberg, 1966.)

pipes have resonance frequencies that are more harmonic than wide ones, and, as a sequel, they produce a greater number of strong spectrum harmonics at low frequencies. Conversely, the spectrum of wide pipes is dominated by the lowest harmonics, and often it is only the fundamental that is really strong. This is illustrated in Figure 5.18, which shows spectrograms of two organ pipes of different widths. In the spectrum a noise contour can also be observed. Its origin is the noise generated by the turbulence that occurs when the airstream is forced through the narrow slit, forming it to an air reed.

This effect of the resonance frequency harmonicity on the spectrum is relevant not only to organ pipes. When all the side holes in a flute are closed, its resonator is a rather narrow tube, because the diameter/length ratio is comparatively small. However, when all holes are open, the resonator is a much wider tube. The consequence is that there are a good number of strong overtones in the spectrum of low flute tones, whereas for higher flute tones, the overtones are weaker and the fundamental more dominating. This causes a timbre difference that is easy to perceive if one really pays attention to it: The lowest tones sound much more rough than the higher ones, which sound more round and smooth.

Air reed instruments often have a typical "chiff" sound starting the tone onset. The reason is that these instruments start to oscillate on an overtone for a short time before the fundamental has had time to get started. The chiff sound makes the tone well-suited for performances of polyphonic music, because it helps the ear to track what the individual voices are doing. The chiff sounds were almost eliminated in the organ building style predominating between 1870 and 1940, an era when harmonies and timbres were more important in the typical organ music repertoire than counterpoint. The trick was to provide the edges in the slit with small dents, or nicks, which made the airstream less turbulent and thus reduced the noise. The onset is comparatively slow, particularly for the bass pipes. As a rule of thumb, a pipe needs about 20 periods to build up a stationary spectrum, which for bass pipes is obviously a considerable amount of duration. Thus onset is slow for low-pitched tones produced in air reed instruments.

In these instruments, the blowing pressure affects fundamental frequency, so that a higher pressure increases the pitch. In the flute, the player can compensate for this by more or less covering the embouchure, thus tuning the resonance frequencies of the resonator. In the organ, the pressure is given once and for all by the weights placed on the bellows. In the recorder, the effect is noticeable: An increased pressure increases the fundamental frequency and this restricts the dynamics of this instrument.

CHAPTER 6

String Instruments

I. INTRODUCTION

The preceding chapter dealt with wind instruments, i.e., music instruments that convert an airstream into sound. We saw that most wind instruments work with a resonator consisting of an air column. There are certainly a number of instruments that do not use these means, and a good deal of these use strings for the sound generation.

Strings can be excited to vibration in different ways. In some instruments the string is struck with some kind of device. Such instruments can be called *instruments with struck strings.* The piano belongs to this type of instrument. In other instruments the string is plucked, i.e., it is pulled out of its equilibrium and then released. Such instruments are referred to as *instruments with plucked strings.* Harpsichord and guitar belong to this group. In a third group the string is excited by a bow, as in the violin family, and this group is referred to as *instruments with bowed strings.*

II. THE STRING

A string can be regarded as a kind of resonator. As in other resonators, the resonance frequencies are determined by the resonator dimensions. If we recall the shape of a piano, it is easy to realize the relationship between dimensions and frequencies. Bass strings are long: Great string lengths are needed for low resonance frequencies. Further, they are thick and spun with a copper wire, so that the string mass is increased: A great string mass contributes to low resonance frequencies. Finally, the pitch rises if the string tension is raised: An increased string tension raises the resonance frequencies.

If we call the string length l, its mass per length unit m, and its tension T, the following formula applies for calculation of the west resonance frequency f_1

$$f_1 = \frac{1}{2l} \cdot \sqrt{T/m} \qquad (6.1)$$

Figure 6.1 shows the envelope for the displacement of the string at the four lowest resonance frequencies under ideal conditions. As the string is clamped at both ends, it is almost impossible to achieve any vibration in these points. A frequency with a wavelength such that it provides displacement nodes at the clamped ends must obviously fit the string best; a clamped end will not move. In other words, the string is a half-wavelength resonator and is, in this respect, similar to a closed-closed tube resonator (no motion at the ends). The resonance frequencies f_n constitute a very nearly, although not perfectly, harmonic series:

$$f_n = \frac{n}{2l} \cdot \sqrt{T/m} \qquad (n = 1, 2, 3, 4, \ldots) \qquad (6.2)$$

The reason for the small departures from harmonicity is that, in reality, a string is far from infinitely flexible, as an ideal string is supposed to be. Rather, a string tends to behave somewhat like a rod. This is particularly true for the stiff strings in a piano.

How a string moves when it is excited depends on the nature of the excitation. For example, the motion is very different for struck and bowed strings. A bowed string divides itself into two straight segments that constitute a certain angle with each other, as we will see soon. This angle travels at a very high speed around the

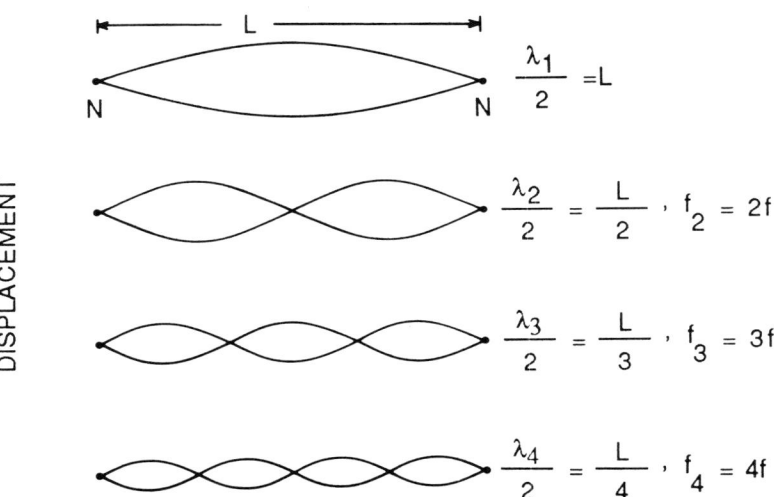

Figure 6.1 Envelopes for the string displacement at the four lowest resonance frequencies under ideal conditions. The string is clamped at both ends, so displacement nodes (N) are found at the ends.

II. The String

lens-shaped contour, which we can see with the naked eye on a vibrating bowed string. In a struck string, there is a little pulse or dent that dashes back and forth along the string. We will return to these matters.

In Chapter 2 the relation was presented between the speed of sound propagation c, the wavelength λ, and the frequency f ($f = c/\lambda$). That equation applies also to strings. As the string is a half-wavelength resonator, $n \cdot \lambda = 2l$. We can now see that the term $\sqrt{T/m}$ in Eq. (6.1) must correspond to the speed of sound wave propagation in the string. We also note that the tension and the mass affect the speed of this wave propagation. The speed concerned in this case is that of the angle in the case of the bowed string and that of the dent in the struck string. This speed will be referred to as the *speed of wave propagation*.

There is a principal difference between tube resonators on the one hand and strings (and also rods) on the other. In tubes the resonator is an enclosed air column, whereas in strings it consists of metal or other materials. As the speed of wave propagation is affected by both the mass and the tension in the string, it is not necessary to vary the string length as widely as in the case of tube resonators, where the speed of sound propagation is constant. A half-wavelength tube resonator with its lowest resonance at 27.5 Hz, which corresponds to the pitch of A0 (the lowest A in a grand piano), requires a half-wave resonator tube length of about 6m. If the lowest string in a piano had to be that long, pianos would have been hard to keep as a piece of furniture, both as grand and upright!

We can get an idea of the order of magnitude of the speed of wave propagation in a piano string by the following example. Suppose that the string length for the pitch of A4 in a piano is 0.4m. According to Eq.(6.1)

$$440 = \frac{1}{2 \cdot 0.4} \sqrt{T/m}$$

where $\sqrt{T/m}$ is the speed of wave propagation in the string. Thus, it equals

$$440 \cdot 0.8 = 352 \text{ m/sec}$$

or similar to the speed of sound. However, the speed of wave propation varies from bass to treble in a piano. The string length for the pitch of A0 (276.5 Hz) may be about 1.32 m, and that for the pitch of A7 (3520 Hz) only 0.06 m in a medium-size instrument. This means that the speed of wave propagation is

$$c = 27.5 \cdot 1.32 \cdot 2 = 72.6 \text{ m/sec} \quad \text{and}$$

$$c = 3520 \cdot 0.06 \cdot 2 = 422.4 \text{ m/sec, respectively.}$$

If we turn to the strings of bowed instruments, we find similar relations. A double bass with the lowest pitch at E1 (41.3 Hz) has a string length of approximately 1.1m. This gives a speed of wave propagation of 90.9 m/sec, whereas a violin with the lowest pitch at G3 (196 Hz) has the string length of 0.327 m, corresponding to a speed of wave propagation of 128 m/sec.

This variation of the speed of wave propagation from bass to treble is a very good idea. The resonance frequency of the string is determined by the length, thickness, and tension of the string. By varying the tension a bit, excessive variation of string length can be avoided.

These speeds of wave propagation in strings are quite different from the values shown in Table 2.1, which concerned the propagation of pressure waves in various media. In strings, however, it is the speed of wave propagation, i.e., the speed of the string deformation, that matters, as mentioned.

The strings course between two supports, and there is a device allowing variation of the tension, as we know, so that the string can be tuned. The string runs over a bridge. This bridge has a mechanical contact with a secondary resonator of some kind, a *sound board* or a *resonance box*. The string vibrations are transmitted to this secondary resonator and from there to the air outside the instrument.

The task of the secondary resonator is, among other things, to amplify the sound generated by the string. This task is extremely important. Thus, if the sound board or resonance box is removed, the instrument becomes almost completely silent, no matter how hard it is played. All of us who have experienced a power failure during an electric guitar concert have experienced this truth in a very memorable way.

For practice and teaching purposes, both pianos and violins have been constructed that lack sound board and resonance box. Playing on a violin that lacks a resonance box is definitely a truly private matter between the player and the instrument: Almost nothing can be heard that can disturb another person.

In some pianos the sound board has been substituted by electronic circuitry providing feedback to the player via earphones; these instruments have been useful in elementary class teaching. All students are seated in the same room with one instrument each, listening to their playing over earphones. The sound of the other instruments is so faint that it does not disturb. The teacher has a switchboard allowing him or her to listen to any student. In this way one teacher can teach several students simultaneously. For more advanced teaching, however, a continuous personal contact between student and teacher would be indispensable.

How come that almost nothing can be heard in the absence of a sound board or a resonance box? The surface area of the string is vanishingly small. It is not easy to achieve motion in a medium with a small tool. Just try to stir your coffee with a needle; the needle is unable to come to grips with the liquid. To move a sufficient number of air particles, i.e., to produce a loud sound, it is necessary that the vigorous string vibrations taking place over a small surface be transformed to smaller vibrations distributed over a larger area. This is arranged by mechanically coupling the string vibrations to the large surface of the resonance box or the sound board.

We realize that string instruments can be regarded as possessing more than one single resonator. If there is only a sound board, there are two resonators. In instruments with a resonance box, there are three: the string system, the air enclosed in the resonance box, and the walls of the resonance box.

III. INSTRUMENTS WITH STRUCK STRINGS

Struck strings are found both in the piano and in the clavichord. The strings are stretched over a frame of iron or wood. Pianos and grand pianos are not constructed in exactly the same way, but as the grand piano is the superior version, we will henceforth consider this type in the first place.

Below the iron frame of the piano there is a wooden board, about 1 cm thick, the *sound board*. It is strengthened by wooden ribs. At the far end the strings course over a bridge glued on this sound board. Because the board is slightly vaulted upward, it presses the bridge up against the strings. In this way, the string vibrations are transmitted to the sound board. Thus, the vaulting is crucial to the radiated sound. The vaulting tends to disappear or decrease by aging and inappropriate treatment (e.g., if the instrument suffers from great and quick changes of temperature or air moisture). This risk is imminent if, for example, a piano is placed near a cold wall. It is difficult to restore the timbre of a piano with a flattened sound board.

The sound source of an instrument with struck strings can be said to be the blow of the hammer against the string. As mentioned, this introduces a little dent that dashes away along the string in both directions and that is reflected at the string ends. As a result, the string starts vibrating on its resonance frequencies. The sound generated is therefore constituted by partials, the frequencies of which correspond to those of the string resonances.

The point where the hammer hits the string has a great significance for the timbre. No spectrum partial can be generated that has a node in that point, because the hammer forces the string to depart from the equilibrium in that point. Thus, if the hammer would hit the string at its midpoint, the missing partials would be the second, fourth, sixth, etc. (i.e., all even-numbered partials). Similarly, if the strike point would be at one-third of the string length, every third partial would be missing.

The situation is entirely different in the clavichord. There the strike is exerted by a metal tongue situated at the far end of the key. This tongue remains in contact with the string as long as the key is pressed. Therefore, the tongue provides the end point of the vibrating string by enforcing a displacement node in this point.

In both piano and clavichord the string vibrations are transmitted to the sound board, as mentioned. Those partials that match a resonance of this board are radiated with a particularly high amplitude. However, there is an immensely great number of vibration modes in a plate of these dimensions, so the resonance frequencies are densely distributed along the frequency scale. The resonance frequencies depend on a great number of factors, and among them, the placement of the ribs and the mechanical properties of the wood are particularly important. A crack may give rise to a jingling sound, probably because the loose edges of the board scrape against each other, almost as the wings of a cricket. Also, the separate parts of a split sound board may vibrate in counterphase, so that the corresponding partials are reduced in amplitude.

In the piano there are two or three strings per key, except for the lowest bass. The hammers, covered with thick felt, hit the strings near their ends. The relatively soft hammer causes a rather smooth dent, somewhat similar to a part of a sine wave. For this reason the source is much less rich in overtones in a normal piano than in instruments with harder hammers such as in tinny, old pianos, in which the hammers are furnished with thumbtacks. Also, if the felt becomes harder due to much playing, the timbre changes and becomes "harder," more rich in overtones. This can be cured by a skilled piano technician, who softens the felt by perforating its upper flanks with needles. This can soften the timbre considerably.

As suggested above, the conditions for a harmonic series of resonance frequencies are not completely fulfilled in piano strings. This is another factor of great timbral relevance, as the spectrum partials correspond to the string resonances. Very unharmonic resonance frequencies lead to an "impure," uneasy timbre.

The harmonicity depends on several factors. One factor is that the piano string is made out of a very hard material, and hence it is by no means infinitely flexible as an ideal string. Rather it tends to behave somewhat like a rod, as mentioned. This applies to all strings but particularly to the ones in the treble. Further, the ends are not inflexible as mountain rocks, but vibrate.

For these reasons, the frequencies of the spectrum partials appear slightly farther away from each other than in a perfectly harmonic series. Instead of, say, 110, 220, 330, 440, etc., Hz, they may appear at 110, 220.3, 330.7, 441.4, etc., Hz. The degree of departure from a harmonic series is called *inharmonicity*. It is generally measured in terms of a coefficient defined as the ratio between the departure from the harmonic value in cents and the squared number of the partial. Thus, if the departure of the third partial is 18 cents, its inharmonicity will be $18/3^2 = 2$. The inharmonicity coefficient b is proportional to the string diameter d and inversely proportional to the square of the frequency f and to the string length raised to the power of 4:

$$b \sim \frac{d^2}{f^2 \cdot l^4} \tag{6.3}$$

The equation reveals a good reason for using strings spun with copper wire in the piano bass; this adds a great deal of mass without increasing its stiffness to the same extent as an equivalent thickening of the string. The effect therefore is a low resonance frequency with a comparatively low inharmonicity.

The significance of the stiffness to the inharmonicity implies that the inharmo-

III. Instruments with Struck Strings

nicity also depends on the string material. Different string materials thus give different timbres. If the strings in an instrument are replaced by strings of a different material, the timbre will change.

It is interesting that the inharmonicity depends on the squared ratio between the string diameter and the frequency. If a piano is not tuned often enough, its tuning generally is flattened. In other words, the frequency drops, whereas the string diameter remains the same, of course. The ratio between the string diameter and the frequency then increases, and according to Eq. (6.3), this will cause the inharmonicity to increase.

The logarithm of the inharmonicity rises approximately linearly with the squared number of the partial, as shown in Figure 6.2. The great significance of the string length implies that a 20% length increase would reduce the inharmonicity by 50% in the treble. However, the tension of the string would have to be increased simultaneously. Large-sized, high-quality grand pianos have a small and evenly increasing inharmonicity (cf. the curve marked 1 in the figure). The inharmonicity is greater in smaller pianos with shorter strings (cf. curve 3 in the figure). What is gained in terms of a reduced instrument size has to be paid for in terms of timbre because of the influence of string length on inharmonicity.

The inharmonicity implies that the spectrum partials do not form perfectly pure intervals. In our example above, we assumed the second partial to be 220.3 rather than 220 Hz. Thus, the octave between the two lowest partials is slightly too wide. Figure 6.3 shows two periods and the associated spectrum for the tone C4 played on a piano at different loudnesses. If we examine the oscillograms carefully, we observe that the first and second periods are by no means exactly alike. This is a consequence of the inharmonicity. In timbral terms the inharmonicity can be heard

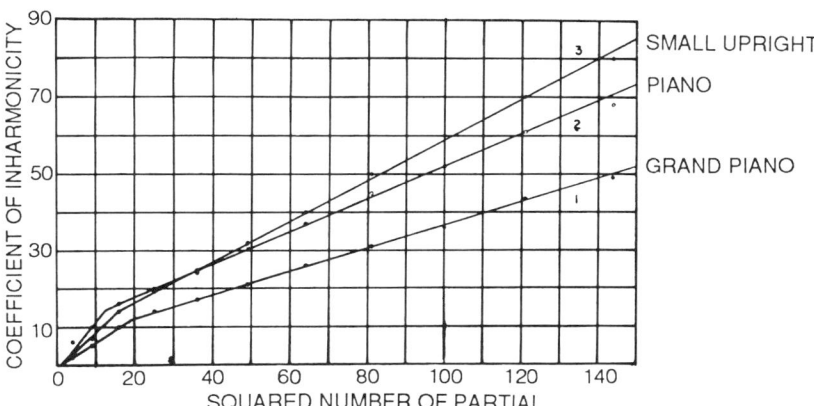

Figure 6.2 Inharmonicity of a bass tone in three pianos of different sizes. (From Schuck and Young, 1943.)

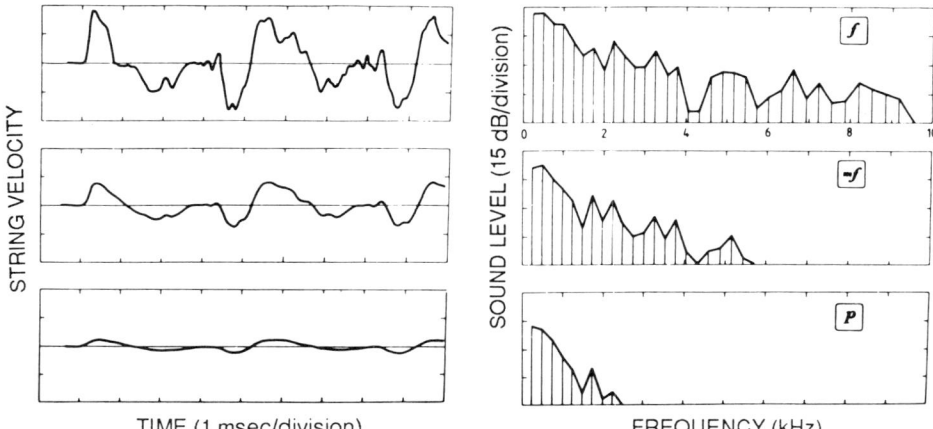

Figure 6.3 Oscillograms (*left*) and spectra (*right*) of three tone onsets played at different dynamic levels on a grand piano. (From Askenfelt and Jansson, 1988.)

as more or less faint beats. Strike one single string (attenuate the neighbor strings, for example, with your fingers!) on a piano and then listen very carefully and you will hear pulsating changes of the timbre.

As was briefly mentioned in Chapter 4, the inharmonicity has consequences for the tuning of the piano. In a harmonic spectrum, the separation between adjacent partials equals the fundamental frequency (e.g., 110, 220, 330), but in a piano the partials are a bit more scarce in frequency. Suppose the frequencies of the lowest partials for the tone A4 are 440, 881, 1,323, and 1,766 Hz. The tuner then adjusts the fundamental frequency of the tone A5 so that as few beats as possible occur. This mostly means that he or she chooses the frequency of the second partial of A4, or 881 Hz in our example. This is 2 cents wider than a mathematically pure octave. As inharmonicity increases with fundamental frequency, also the octaves become more and more "stretched." The highest note on a piano may be no less than 35 cents higher than what it would be if mathematically pure octaves were used.

A similar reasoning applies to the bass. If the note of A3 is to give a beat-free octave with the note of A4, its second partial should agree with the first partial of A4. As the partials of A3 are also slightly inharmonic, this means that a frequency lower than 220 Hz has to be selected, perhaps 219.5 Hz, which is 4 cents below the mathematically "correct" value. Also in this case, the deviations increase with increasing distance to the starting point, and the lowest notes in a grand piano may be no less than 40 cents flat as compared with what they would be if mathematically pure octaves had been used. The average deviations from the mathematically correct, equally tempered tuning observed in grand pianos is shown in Figure 6.4.

III. Instruments with Struck Strings

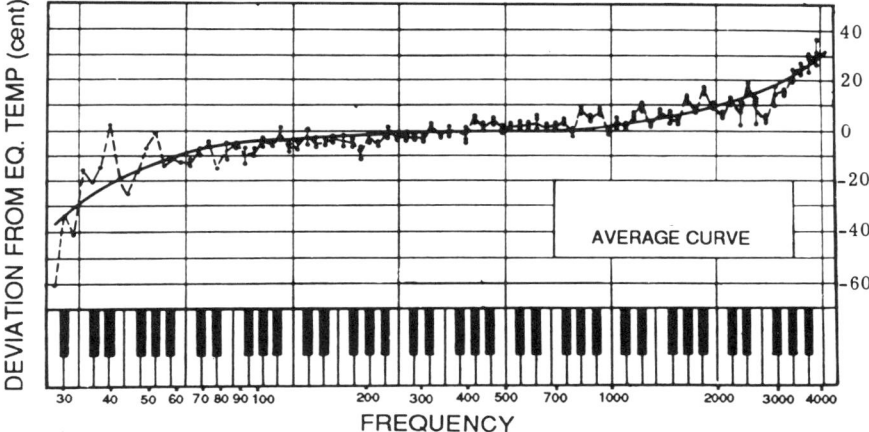

Figure 6.4 Heavy curve shows average deviation from the frequency values of the equally tempered tuning in mid-sized grand pianos. Points represent measurements on single strings in an individual piano tuned by a fine tuner. (From Martin and Ward, 1961.)

Thus, the inharmonicity imposes the need for stretching the octaves in the piano to minimize beats. As explained in Chapter 4, the octaves need to be slightly stretched to be perceived as pure.

There is generally a fair degree of disagreement between physicists and pianists regarding the pianist's possibilities to affect the timbre by means of the touch. The problem is that the hammer has no contact with the key when it hits the string. As viewed from a physical standpoint, the only possibility is to vary the hammer velocity. This implies that loudness and timbre are interdependent: If one wants to change the timbre for an individual key, the only way to do it is by changing its loudness.

Pianists normally do not agree with this view. They know that they can produce a number of different timbres keeping the loudness constant by varying the touch. Could this be a mistake? The late music acoustician A. H. Benade argued that musicians are never wrong, because they create the reality that music acoustics tries to explain. Also, they are professional experts in shaping and listening to sound.

One possibility is the following. According to more recent analyses the hammer and its shank can play a number of funny tricks on the way up to the collision with the string, depending on how the key is struck. The hammer is a mass placed at the end of a thin wooden stick that is somewhat elastic, so this system has resonances that can be excited. Thus, for a special type of touch, the hammer sways up and down. Therefore, the hammer can actually scratch along the string during the time when it is in contact with the string. This may affect the timbre.

Further, if the finger hits the key from the air, the key does not move at a constant rate down to the bottom stop. Rather the key speed varies. Probably there is a vigorous reaction of the mechanism that may give the key an extra push. We do not know very much about the effects of these phenomena. However, it does not seem physically impossible that different ways of pressing the key give rise to different types of collision between the hammer and the string, which results in different timbre for the same loudness.

One should also realize that "touch" is not a crisp clear concept. One might suspect that it sometimes may mean, for example, "the art of selecting such loudnesses for the various tones in a chord that the intended timbre of a chord is created." Further, when it comes to the playing of the solo voice in a piece of music, "touch" may mean the art of achieving a similar timbre for neighboring pitches. Also, the precise timing of the individual tones in the chord are probably quite important to the perceived timbre. For example, an early arrival of one tone tends to make this tone stand out from the other chord tones to give it the function of a solo part. We will return to such effects in Chapter 10.

The characteristic feature of a piano tone, that is, what reveals to us that the tone originates from a piano, is the typical amplitude pattern. Figure 6.5 shows a typical example. The pattern consists of two parts. In the first part, the amplitude decays very quickly, whereas in the second part it dies away very slowly. This reflects the way the strings are vibrating. During the first part, the strings vibrate up and down,

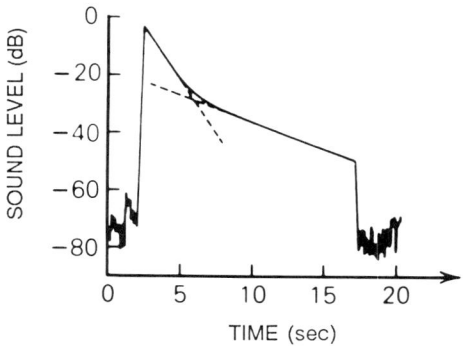

Figure 6.5 Typical amplitude pattern for a piano tone. During the first part, when the string is vibrating up and down, the amplitude decreases quickly, and during the second part, when the string vibrates sideways, parallel to the sound board, the amplitude decreases slower. (From Weinreich, 1977.)

III. Instruments with Struck Strings

i.e., away from and toward the sound board as we might expect, in view of the striking of the hammer. During this phase, the string exports lots of energy down to the sound board, because the bridge has no great objection against vibrating this way. However, the plane of the string motion soon rotates, so that the string starts to vibrate in parallel with the sound board. It is obviously much harder to convince the sound board to vibrate in this direction. The string therefore loses less energy to the sound board and hence retains its amplitude for a longer time.

The amplitude pattern is modified by the *una corda* pedal. The effect of pressing it is that the hammer just hits two of the three strings belonging to the key. The third string brought to vibration via the bridge vibration induced by the hit strings starts to vibrate parallel to the sound board and so accelerates the vaulting of the vibration of the two struck strings. The effects on the amplitude pattern can almost be guessed. The non-struck string shortens the first part of the amplitude pattern and increases the amplitude of the second part. The attack becomes smoother, thus making the perceived tone softer.

It is not only the change in the decay rate that is typical for a piano tone. Up to now we have considered the characteristics of the string vibrations. Another very typical property is found in the radiated sound. The very first fraction of the sound of a piano tone is constituted by a high-frequency percussion sound, the so-called forerunner shown in Figure 6.6. As a matter of fact, a synthetic piano tone sounds more like a harp tone if this forerunner is missing. The forerunner seems to have an interesting origin. The speed of the shock wave propagation in a medium is greater for high frequencies than for low frequencies. Therefore the high-frequency components of the shock sound are quicker to reach out into the air than the lower ones.

The clavichord differs from the piano with regard to the excitation mechanism as well as the string system. Instead of the hammer, a metal tongue is fastened to the far end of the key that is a double-armed lever. When the key is pressed, the tongue rushes upward and hits the string with its sharp edge. The excitation is much richer in high overtones than in the case of the piano because the dent in the string is sharper.

The tongue keeps pressing the string upward as long as the key is pressed down, as mentioned before. Therefore, the tongue constitutes the far end termination of the string. In other words, the point of contact between string and tongue must be chosen such that the resulting string length corresponds to what the pitch required. The part of the string that is beyond the point of contact does not participate in the vibration; it is damped with felt strips. It is the other part of the string that is vibrating. The sound is very soft. As the fundamental frequency is dependent on the contact point between tongue and string, the same string may be used for two tones, which, however, cannot be played simultaneously in such cases.

The tension in the vibrating part of the string is affected by the force by which

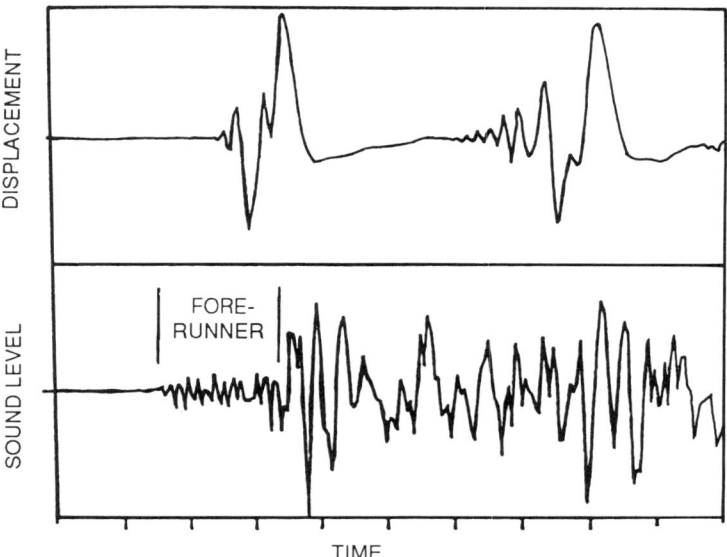

Figure 6.6 Oscillogram of the onset of a piano tone as recorded in terms of string vibration (*above*) and in terms of the sound near the instrument (*below*). Onset starts with a high-frequency percussion sound, the forerunner, which is an important characteristic for piano tones. The time scale is 7.8 msec per division. (From Podlesack and Lee, 1988.)

the key is pressed. As the tension affects the resonance frequency, it is possible to vary the tuning by changing the pressure on the key. Thus, by varying the key pressure rhythmically, a vibrato-like modulation of the fundamental frequency can be produced, which is called *Bebung* in German, "quake."

IV. Instruments with Plucked Strings

In this type of instrument the string vibrations are caused when the string is lifted by means of the finger or a pick and then released. The result is a sound source with a great number of strong high harmonics.

The wave propagation in plucked strings is different from that in a struck string. The string divides itself into three straight-line sections coursing between two angles, as shown in Figure 6.7. These angles dash around along the lens-shaped contour, which we can observe with the naked eye when the string is vibrating.

The harpsichord has a stringing similar to that of a piano, although the strings of the harpsichord are much thinner, less tense, and mostly longer. Further, the

IV. Instruments with Plucked Strings 153

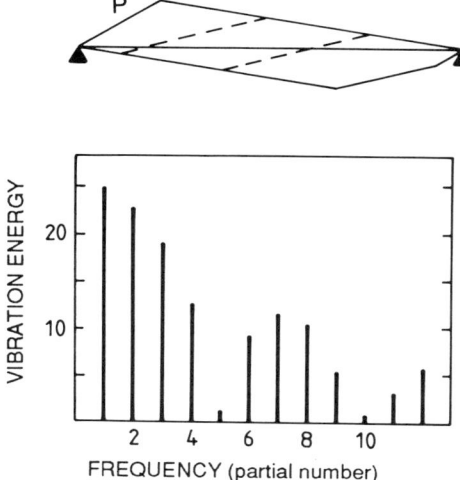

Figure 6.7 *Upper graph* illustrates string deformations during the first half period. The string is plucked in the point P at one-fifth of its length. Two kinks rush away from P in opposite directions and then reflect at the clamped string ends. *Lower graph* shows the corresponding spectrum. (From Hall, 1980.)

harpsichord has several registers, each corresponding to a complete set of strings. Some registers are tuned one octave higher than normal, so-called 4-foot registers versus the normal 8-foot register.* By varying the string material, timbral characteristics can be also created between registers with the same tuning.

When the key is pressed, the string is plucked by a small thorn made of leather or nylon. Regarding the significance of the point of excitation to the timbre, the same applies as in the case of the piano: an antinode is enforced in the point of excitation. Partials with a node in this point cannot be generated.

The sound radiation is enhanced by a sound board, which in some instruments consists of a wooden box with a heavy bottom. In this manner an extra acoustic resonator is obtained apart from the mechanical one provided by the regular sound board. Probably, this acoustic resonator adds some low-frequency resonances, which affect the timbre by making it more "sonorous."

In other plucked instruments such as the lute, guitar, banjo, and mandolin, the

---

*These terms, which may appear somewhat exotic, originate from the world of the organ. The lowest key on a normal organ key board gives the pitch of C2 (approximately 65 Hz). It corresponds to a pipe of about 8-foot length, provided the pipe is open in the upper end. Hence, harpsicord registers having this tuning are also called 8-foot registers. If a register sounds one octave higher or lower, the pipe length in the organ is half (i.e., 4 foot) and double (i.e., 16 foot), respectively.

strings are plucked by the fingernails or by a plectrum. The sound radiation is improved by a resonator box in terms of a resonance box with thin wooden walls and a sound hole. Such instruments contain three resonators, as mentioned: the string, the resonance box, and the air contained in this box. The characteristics of the radiated sound depends on several factors: the plucking, the string, and the resonances of the air volume.

The electric guitar lacks both the acoustic and the mechanical resonators represented by the sound box in the traditional guitar. The vibrations of each string are caught by a small electric magnet device, a pickup. Its location determines what partials will be missing in the spectrum: All partials with a node in that point will be missing. The pickup is mostly quite close to the bridge, and in this way a timbre rich in high harmonics is produced. On some electric guitars the pickups can be slid along the strings so that also more soft timbres--lacking more harmonics--can be obtained. Some more ambitious electrical guitars have several sets of pickups that the player can chose between.

The signals recorded by the pickups are processed by amplifiers, and their characteristics are very significant to the resulting timbre. The amplifier thus replaces the resonance box of the traditional guitar. Many players consider amplifiers equipped with vacuum tubes the best. The decisive difference between amplifiers with vacuum tubes and transistors would be that the distortion of the signal, which provides an important timbral effect in these instruments, can be varied more gradually in a vacuum tube amplifier whereas it occurs more suddenly in a transistor amplifier. Gradual changes of sound quality are probably easier to use for the purpose of musical expression.

V. Instruments with Bowed Strings

The violin is an instrument that has been surrounded by a dense mysticism for a very long time. Lately, the mysticism has evaporated to a significant extent thanks to a devoted research activity spent by two leading persons, Carleen M. Hutchins, of the United States, who shares her time between building and researching bowed instruments, and the Swedish music acoustician Erik Jansson. Without the knowledge that has emerged from their research, much of what follows would have been missing.

In instruments in which a bow is used to excite the strings, the source consists of a periodic signal. The horsehair is treated with rosin to produce friction against the string. The effect of this friction is that the bow grabs the string and pulls it away a little from its equilibrium until the tension overrides it. Then, the string slips back, past the equilibrium, until it again sticks to the bow and joins its motion. The situation can be described somewhat more in detail in the following way.

V. Instruments with Bowed Strings 155

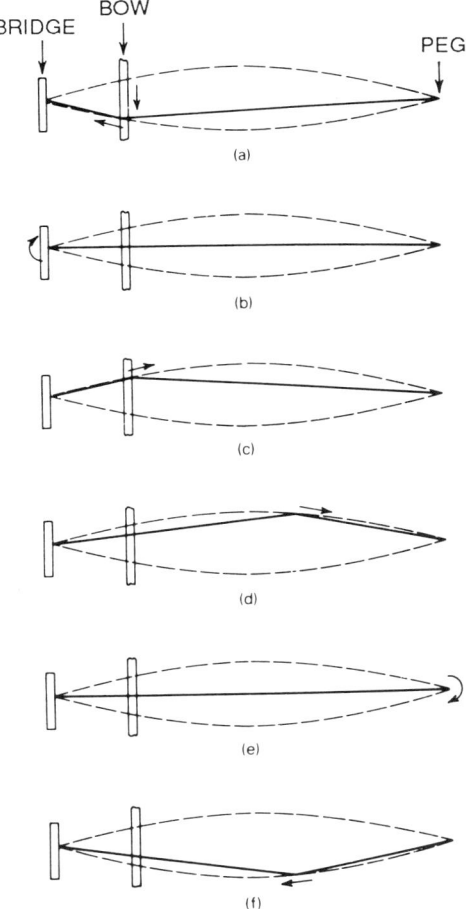

Figure 6.8 Series of snapshots illustrating the vibration pattern of a bowed string. The bow, which is moved downward in the figure, has grabbed the string in (a). In (b) and (c), the string slips upward, and in (d), (e), and (f), the bow has caught the string again. (From Hall, 1980.)

The shape of the bowed string invariably assumes the shape of two straight lines meeting under a certain angle, as mentioned before (see Figure 6.8). This angle travels along the string on a lens-shaped course and switches direction at the ends. If one examines the figure carefully, one can see that the string changes the direction of travel when the angle passes the bow after having passed the bridge end. In that very moment, it is particularly easy for the bow to catch the string and pull it, because the string then moves in the same direction as the bow. The string thus joins the bow until next time the angle passes the bow and changes its direction of motion. In that moment the string slips loose from the bow, and then the same pattern is repeated.

The player can affect the string vibration within wide limits. The *bow velocity* is the tool for regulating the amplitude of vibration. This is not surprising, since a great bow velocity must pull the string far away from its equilibrium and thus produce a large vibration amplitude.

The point where the bow excites the string, or the *bowing point,* affects the overtone content of the vibration: the closer to the bridge, the more overtones. The bow enforces an antinode in the bowing point, so partials with a node in that point cannot occur. Hence, the closer the bowing point is to the bridge, the more high frequencies are produced. The bowing point also influences the amplitude.

The player adapts the *bowing force* against the string to the bow velocity and the bowing point. This force, which is sometimes incorrectly referred to as the bowing pressure, also influences the tone quality. This is because the angle, which becomes somewhat smoothed when it passes the end point of the string, is sharpened by the bow: the greater the bowing force, the sharper the angle. This sharpness pays off in terms of high overtones. The bowing force also decides if the bow is able to catch the string, and if it fails, there will be no tone at all, only a scratching sound.

If observed at a given point, the bowed string vibrates according to a saw-tooth curve, as shown in Figure 6.9. The fundamental frequency is determined by the string's lowest resonance frequency except for flageole tones, when a higher resonance is the determinant. One branch of the sawtooth curve is steeper than the other; the closer to the bridge the bowing point, the greater is the asymmetry. Its amplitude is controlled by the bow velocity, as mentioned.

Theoretically a periodic sawtooth waveform corresponds to a spectrum of harmonic partials with both odd- and even-numbered partials. The amplitudes decrease at a rate of about 6 dB per octave, as can be seen in Figure 6.9. However, the vibrations of a bowed string are a bit more complicated than that. The string also rotates and its tension varies with the displacement. The effect is a slight random variation of the frequency. This adds a highly characteristic property to the timbre; if by synthetic means one eliminates nothing but this random variation in a violin tone, the resulting timbre does not possess even a remote similarity with the typical violin sound.

The resonance box of a violin can be regarded as an acoustic-mechanical resonator. The mechanical part of it is the top and bottom plates and the sidewalls. It is excited by the string vibrations that are transmitted by means of the bridge. The string vibrates back and forth in the direction of the bow motion and so rocks the bridge, which, in turn, vibrates the top plate. Below one of the bridge feet a thin pin is placed, the sound post, which is wedged between the top and bottom plates. The transmission is greatest through the bridge foot standing above this sound post. Thus the bridge foot above the sound post rhythmically presses the top

V. Instruments with Bowed Strings

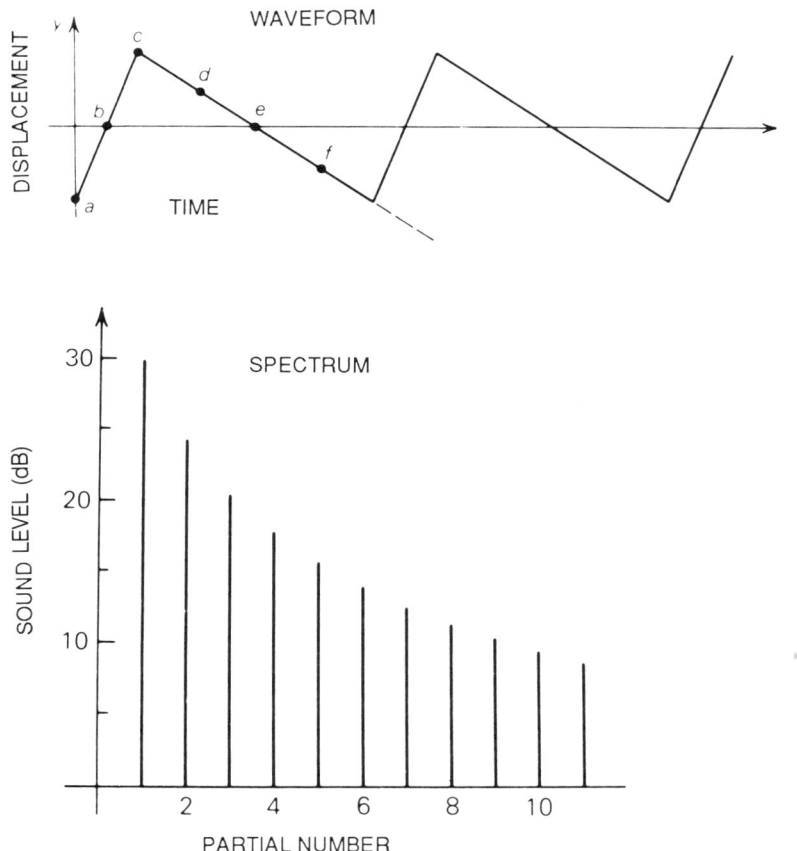

Figure 6.9 Schematic illustration of the waveform (*upper graph*) and spectrum (lower graph) for bowed string vibrations. The letters correspond to those in Figure 6.8. (From Hall, 1980.)

plate down in pace with the bridge and string vibrations. It stamps the string vibrations into the roof of the resonance box.

How the radiated sound turns out also depends on the willingness of the top plate to vibrate at the frequencies proposed by the string vibrations. This willingness can be measured in terms of an *input admittance curve*. Figure 6.10 shows an example of such a curve. It reflects the resonance properties of the entire resonator system.

An input admittance curve thus shows the willingness of vibration as a function of frequency. The input admittance curve recorded at the bridge foot of good instruments shows a typical pattern. It is characterized by four approximately

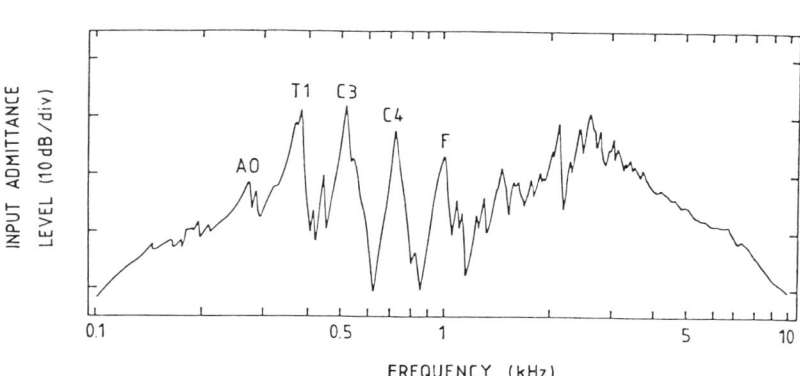

Figure 6.10 Input admittance curve for an excellent violin recorded by rocking the bridge at different frequencies. (From Alonso Moral and Jansson, 1982.)

equidistant peaks of similar amplitudes below 1 kHz. In less perfect instruments, these peaks are of different heights and unevenly distributed along the frequency scale. In good instruments the curve drops above 1 kHz and then again rises to a peak near 3 kHz. This peak is dependent on the bridge, which therefore has a great significance for the end result.

The great significance of the bridge to the timbre is not hard to understand. The bridge really deserves its name; it allows the string vibrations to pass down to the resonance box. Hence, its transmission characteristics must count. Its great importance is illustrated by the considerable timbral effects of adding a mute, the effect of which is merely to add a small mass to the bridge. This affects its vibration transmission characteristics and therefore changes the timbre.

To maintain normal string vibrations, the bridge must possess a certain stability, so that it does not vibrate too vigorously during playing. If the bridge fails to fulfill this demand, e.g., because the top plate vibrates greatly under the bridge foot at a certain frequency due to resonance, the tone becomes unstable. This phenomenon is called the *wolf note* and is a problem in poorly built violins and particularly cellos.

The sound box embraces an air volume that constitutes the acoustical resonator of the instrument. It is excited by, and to some extent also affects, the mechanical system, and thus the string played. The resonance properties of the mechanical system are determined by the mechanical properties of the sound box (i.e., the wood thickness and vaulting). The characteristics of the acoustical system is mainly determined by the shape of the sound box and the f-holes.

We realize that many different factors contribute to determining the resonance properties of the sound box. These are important to the quality of the instrument.

V. Instruments with Bowed Strings

A great deal of research work has been spent on exploring the characteristics typical of high-quality instruments. What is needed for such investigations is a recording revealing how the plates vibrate at different resonances.

By means of laser light, the distribution of vibratory displacement over a plate can be recorded. The trick is to bring the plate into vibration at various frequencies and take a photo by means of a special technique. The result displays the vibration amplitude approximately as a topographical map, i.e., the higher the number of contours shown in the photo, the higher the vibration amplitude. Figure 6.11 shows an example.

A simpler, although less exact way is as follows. One sprinkles the plate with a light, visible powder, e.g., cork filings. Then the plate is brought to vibrate by placing it above a loudspeaker that emits a very loud sine tone with variable frequency. When the loudspeaker emits a resonance frequency of the plate, it starts to vibrate vigorously. Then, the powder assembles along the node contours.

By reading the resulting vibration patterns, one can decide where to thin the plate to change the resonance properties as desired. Many violin makers use this method routinely for measuring and tuning the resonances of the plates before gluing the sound box together.

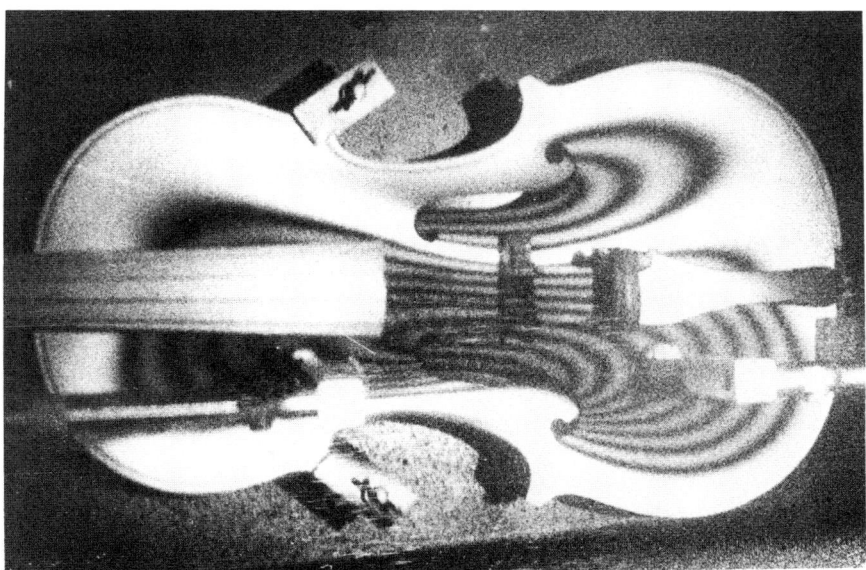

Figure 6.11 Vibration amplitude in a violin top plate at 480 Hz recorded by laser technique. The higher the number of contours surrounding a bright area, the greater the vibration amplitude in that area. One contour corresponds to a vibration amplitude difference of 1/1,000 mm. (From Jansson and Alonso Moral, 1980.)

Particularly the varnish has been surrounded by mysticism at all times. One has assumed that it represents the kernel of the secrets of the classical Italian violin building. This is probably not true. It is possible that the amplitude of the resonance peaks can be adjusted by the varnish, but probably, its main significance is to reduce the wood's sensitivity to changes of the ambient moisture. The effect of this is that the sound box retains its resonance properties irrespective of, for example, weather and climate changes, which is certainly a great advantage. However, this does not make the varnish the kernel of the secrets of the classical Italian art of violin making.

The resonance properties of bowed instruments are regarded as a decisive factor for the quality. At the same time it is dependent on a great number of instrument properties that affect each other in a very complicated system. Against this background it is not hard to understand that the mystery around violin making has been difficult to resolve and that traditional methods are regarded as very important.

CHAPTER 7

Rod and Membrane Instruments

I. ROD INSTRUMENTS

In some instruments, a rod or a membrane is used for sound generation and generally a strike is used for exciting it. Rods are used in the triangles, xylophones, marimbas, celestas, in some music boxes, and in tuning forks, which could also be regarded as the same type of musical instrument. Expanding our terminological generosity somewhat, we may also regard bells and gongs as severely arched membranes and count them as members of the same instrument family. We will call all these instruments *rod* and *membrane instruments*.

In rod instruments the sound is generated by a strike on the rod. Thereby, it starts to sound with a spectrum of overtones corresponding to its resonance frequencies.

Recalling the vibration principles of strings, it appears natural that a rod clamped in one end and free in the other end acts as a kind of quarter-wave resonator. An instrument with this type of resonator is the steel combs that can be found in certain music boxes. If the rod is clamped in the middle, as in the tuning fork, the rod acts rather like a half-wave resonator (i.e., the lowest resonance frequency has a wavelength such that antinodes appear in the free ends and a node in the clamped middle). Figure 7.1 illustrates this situation. Note that the standing wave patterns do not display clear antinodes in the free ends. The displacement amplitude is great, but the free end is not located at the maximum of the sine wave curve describing the amplitude.

The resonance frequencies of a rod are far less harmonic than those of strings and tube resonators. Let us imagine a rod of length l and rectangular cross section with the thickness t. Let us further imagine that the rod is lying on supports such that it is free to vibrate at both ends. The lowest resonance frequency f_1 then is

$$f_1 = 1.028 \frac{t}{l^2} \sqrt{\frac{Y}{D}} \tag{7.1}$$

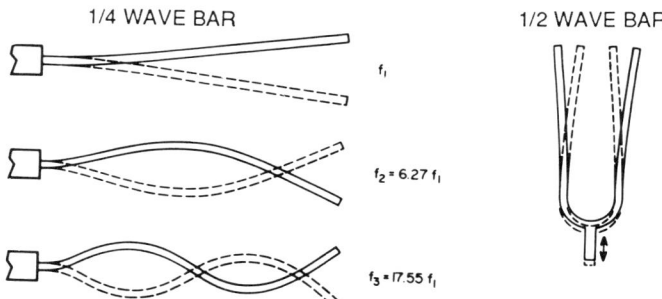

Figure 7.1 *Left figure* shows the three lowest modes of vibration in a rod clamped in one end and free in the other end. *Right figure* represents the amplitude pattern for the lowest vibration mode in a rod clamped in the middle, as in the tuning fork. (From Backus, 1969.)

where Y is a measure of the stiffness, called Young's modulus, and D is the mass density of the rod. If the rod has a circular cross section with the radius r, the lowest resonance appears at

$$f_1 = 1.78 \frac{t}{l^2} \sqrt{\frac{Y}{D}} \tag{7.2}$$

A circular rod thus has its lowest resonance about a major sixth higher than an equally long rod of rectangular cross section. In both these cases, the higher resonance frequencies f_n appear at

$$f_n = 0.441 \, (n + 1/2)^2 \cdot f_1 \tag{7.3}$$

i.e., the series of resonance frequencies is the following:

$$f_1 = 1.00 \cdot f_1$$
$$f_2 = 2.76 \cdot f_1$$
$$f_3 = 5.40 \cdot f_1$$
$$f_4 = 8.93 \cdot f_1$$

If the rod is clamped in one end, the lowest resonance for the rectangular rod is

$$f_1 = 0.162 \frac{t}{l^2} \sqrt{\frac{Y}{D}} \tag{7.4}$$

and for the circular rod

$$f_1 = 0.28 \frac{t}{l^2} \sqrt{\frac{Y}{D}} \tag{7.5}$$

I. Rod Instruments

or again about a major sixth higher. The higher resonances appear at

$$f_n = 0.441(n + 1/2)^2 \cdot f_1 \qquad (7.6)$$

so that the series becomes

$$f_1 = 1.00 \cdot f_1$$
$$f_2 = 6.32 \cdot f_1$$
$$f_3 = 17.55 \cdot f_1$$
$$f_4 = 34.42 \cdot f_1$$

We can see that a doubling of the length lowers the frequency to a quarter, or two octaves instead of one octave, as in strings and tubes.

Note also that the resonance frequencies not even approach a harmonic series. If f_1 corresponds to the pitch of C3, the overtones in a rectangular rod appear at a low Gb4, a sharp F5, and a flat D6. A good deal of the typical "metallic" sound quality of a struck rod originates from this very inharmonic series of overtones.

The higher resonances disappear very quickly in the sound. Therefore, if one succeeds in juggling with the shape of the rod such that its second resonance appears either a duodecime or a double octave above the lowest resonance, the spectrum gains in harmonicity. The modification needed is to make the rod more flexible in the middle of its length, e.g., by making it thinner there. Then, the first resonance drops considerably in frequency. This trick is used for the greater portion of the pitch range of the marimba, where the first resonance lies two octaves below the second. In the treble, however, the need for being that meticulous is not all that great, so there it does not pay off to bother about this arrangement. The middle of xylophone rods are not thinned to the same extent, so the first resonance is only a duodecime below the second one.

If one wants to excite and listen to the resonances in a rod, one must grasp it in a node and strike it in an antinode; it can easily be realized that the rod cannot vibrate on a resonance if struck in a node or clamped in an antinode. In a rod that is free to vibrate in both ends, the node of the first resonance is found at about 20% of the length. Thus, if one wants to enjoy the sound of the first resonance of a rod of 1-m length, one should grasp it about 20 cm from one end. If one holds it in the end, there will not be much of a sound to enjoy, because no resonance has a node in this point.

The rods of the celesta and the music box are fastened in one end and free in the opposite end. In the xylophone, they are placed horizontally on two supports. The celesta and the xylophone are further furnished with extra acoustic resonators, one per rod. Their task is to transform the great rod vibrations over the small rod area to smaller vibrations over a larger area. In this way the radiated sound is amplified considerably, exactly as in string instruments with a sound board or a resonance box. The resonators of the celesta consist of small boxes, those of the xylophone of tubes. The resonance characteristics of these extra resonators, of course, contribute substantially to coloring the radiated sound.

The inharmonicity of the partials cause the sound to change continuously. After all, even the modest inharmonicity of the piano strings is sufficient for causing this effect, as mentioned. The metallic sound is partly due to the quickly decaying inharmonic higher partials, as mentioned. In the case of the tuning fork the higher partials are rather weak, so that a rather sinusoidal like waveform is generated.

II. MEMBRANE INSTRUMENTS

Kettledrum, drum, and tambourine are examples of membrane instruments. The sound is generated by the membrane, which is tensed over a circular frame. The vibrational characteristics of the membrane depart drastically from those typical for pipes, strings, and rods. This is not surprising. It would be hard to imagine that a circular object such as a membrane vibrated in the same way as an oblong object such as an air column or a string. In a rather complex fashion the membrane divides itself into different fields that vibrate simultaneously (see Figure 7.2).

A number of factors determine the locations of these fields. Let us start by considering an ideal membrane, equally flexible and equally tensed in all directions, the tension being independent of the membrane displacement. Under these conditions, the lowest resonance frequency f_1 depends on the membrane radius r, the tension T, and the mass m per square unit:

$$f_1 = \frac{0.38}{r} \sqrt{\frac{T}{m}} \qquad (7.7)$$

In a mylar membrane of a kettledrum, the radius may be 0.3 m, the mass 0.26 kg/m², and the tension 2·10³ N/m. In that case,

$$f_1 = 111 \text{ Hz}$$
$$f_2 = 1.59 \cdot f_1$$
$$f_3 = 2.13 \cdot f_1$$
$$f_4 = 2.29 \cdot f_1$$
$$f_5 = 2.65 \cdot f_1$$

The resonances are inharmonic and rather dense in frequency, much more so than in rods. In the real case there are a number of other significant effects. For instance, the air on both sides of the membrane adds to its real mass.

The resonance frequencies of the membrane depends, among other things, on its tension; kettledrums are tuned by varying the tension of the membrane. The

II. Membrane Instruments

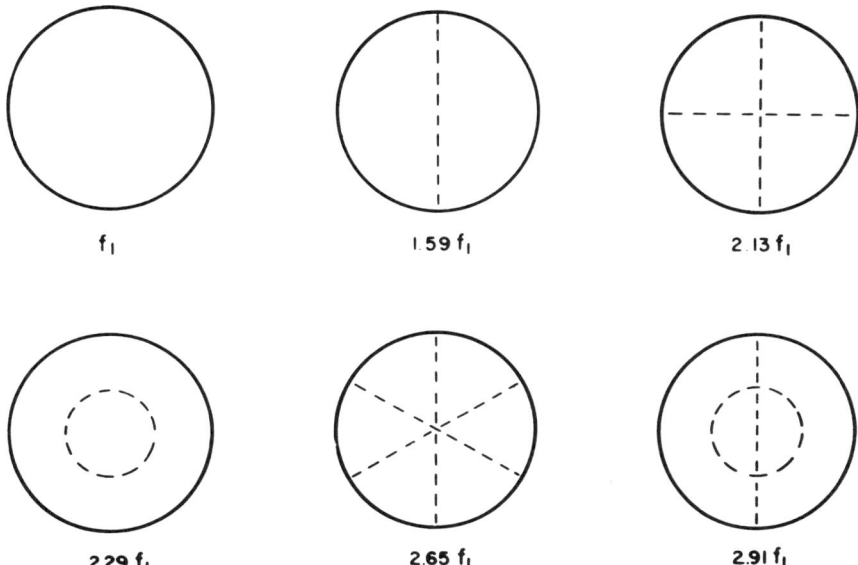

Figure 7.2 Partitions of an ideal membrane into vibrating fields. *Dashed contours* represent the nodal lines, where the membrane does not vibrate. (From Backus, 1969.)

exact way in which the membrane is tensed is also of significance to the tuning of the individual resonance frequencies. For example, in the Indian membrane instrument tabla, the membrane is tuned with a great sophistication. The tension and the mass of the membrane are adjusted according to a complicated system so that almost harmonic resonance frequencies are obtained.

The frequencies of the resonances depend on how the membrane divides itself into vibrating islands, as shown in Figure 7.2. One way is to add mass to certain spots of the membrane. Another way is used in a membrane instrument manufactured and played with great skill in some Caribbean countries, the steel pan. Here the resonance frequencies are tuned by using a very stiff, arched membrane that is made thinner along certain elliptical contours. A number of different sized contours are arranged side by side on the membrane, such that a series of areas are obtained, each possessing its own resonance. The resonances are tuned so that they constitute a scale, and the tone is produced by striking with a mallet in the respective areas. The arching of the individual island is very significant to the pitch obtained from it. The instrument tuning can easily be disturbed by touching the islands too hard.

The recipes for how vibrating membranes are divided by nodal lines also applies for the highly arched membranes in gongs, cymbals, and bells. In a bell the thickness of the membrane is a factor of particularly great importance.

In most membrane instruments, the membrane is tensed over a closed cavity, which thus acts as a secondary resonator. The sound from the struck membrane is reflected in the cavity, the resonance of which contributes to increasing the vibration amplitude. For this reason kettles of different sizes are needed for differently tuned kettledrums. These instruments have a hole in the bottom of the kettle. The partial corresponding to the lowest resonance is the only resonance that makes a strong effort to drive air particles through that hole. Therefore this resonance decays quickly as the friction in the table steals energy particularly at low frequencies.

Another task of the cavity is to separate the upper and the lower membrane surfaces. This is important. If these surfaces are not separated, a strike will cause a rarefaction of the air on the strike side and an increased density below the membrane. Then, the air would simply slink round the edge such that the air missing on the upper side will be imported from the lower side. This would cancel much of the sound wave. Placing the membrane over the cavity therefore enhances the sound obtained from a strike.

In some instruments the cavity is open, as in the tambourine. This trick is also used in some drums, in which one of the drumheads is eliminated. This considerably attenuates the radiated sound, because the open end of the cavity reflects sound much less efficiently than a closed end and thus contributes far less to the vibration amplitude of the membrane. Also the separation of front and back side is, of course, less efficient.

When membranes are struck, the vibrational response and hence the sound depends also on the point where the mallet hits the membrane. In this case the same principles apply as for strings; only those resonances are excited that have an antinode at the striking point. If a kettledrum is struck in the center, only few resonances are excited, as can be realized from Figure 7.2. Remember also that the first resonance is attenuated by the center hole in the kettledrum. The decay of this resonance is therefore very quick, and the resulting sound is somewhat similar to a muffled thud. Often the membrane is struck at half or two-thirds of the membrane radius. This excites resonances 2, 3, 5, 7, 10, ..., and in this way a rather clear pitch is obtained.

The mechanical properties of the mallet are of great significance to the sound. A soft mallet creates a dent of lesser depth over a wider area than a harder mallet. As a result, the dent hits simultaneously both nodes and antinodes of the high-frequency resonances, so these will not get excited. A harder tool will make a more precise and deeper dent over a narrower area and so excites also higher resonances. Therefore the sound contains more high-frequency components. Another factor of relevance is the length of time during which the mallet has contact with the membrane. If this time is too long, all resonances with an antinode in the striking point are attenuated, so that the decay becomes faster.

The spectrum of the radiated sound is very inharmonic in many membrane instruments. Therefore the timbre is constantly changing, exactly as in the rod instruments. These inharmonic partials may also cause difficulties in perceiving a clear pitch in the sound, particularly if the membrane is not correctly strung over the kettle.

III. BELLS AND GONGS

A membrane can be more or less arched. The steel pan, for example, has a slightly vaulted membrane. Gongs are a bit more vaulted, and bells can be regarded as strongly arched membranes. Thus, they contain resonances similar to those of a membrane. However, the resonance frequencies are very dependent on the thickness which varies considerably.

There are several types of bells: the classical church bell, carillon bells, handbells, and in Asian cultures, there are other types of bells with quite special shapes. The bell is one of the oldest tools for making sound and it is sometimes used for musical purposes such as in carillons. Also, bells are used in some operas, e.g., Richard Wagner's *Parsifal*; however, real bells are rarely used in the orchestral pit in opera houses for obvious reasons: they are too loud and too big. Instead, electronic means or orchestral chimes are used.

The vibrations at the various resonance or mode frequencies can be described in the same manner as membranes. The surface is divided into vibrating areas separated by nodal lines. The modes of bells are classified by means of the location of the nodal lines specified in terms of two numbers within parentheses, e.g., (2,3). The first number refers to the number of complete nodal meridians over the top of the bell and the second number refers to the number of nodal circles around the bell. For example, mode (3,1) produces three nodal meridians and one nodal circle. It is possible to tune these modes by making the bell thinner or thicker in different places.

The modes of church bells are generally tuned in a particular way, so that the frequencies of the lowest partials have the ratio 1:2:2.4:3:4. If we translate this to intervals the result is octave, minor third, fifth, octave. The tuning of these partials is realized by originally making the bell a bit too thick and then thinning the wall material at particular heights. When this tuning has been successfully completed, some additional partials also become harmonic, and so the sound is perceived as somewhat harmonic, possessing a reasonably clear pitch. Table 7.1 lists some of the modes, their names, and ideal frequency ratios in just tuning. In reality, the higher mode frequencies are higher than the ideal values. Also shown are decay times for these modes as measured in a bell of 70 cm diameter.

When struck by the clapper, a sharp, metallic-sounding bang emerges consisting of a great number of inharmonic partials that decay quickly. After a short while, the lower partials become dominating in the sound, and the apparent pitch is mostly that of the second partial, or the strike note one octave above the lowest hum note. After a while, only the hum tone remains, an octave below the strike note.

Table 7.1 Modes and Partials in an Ideal Church Bell[a]

Mode	Name	Ratio to strike	Interval to strike	Decay time[b] (s)
(2,0)	Hum	0.5	Octave	52
(2,1)	Strike	1.0	Prime	16
(3,1)	Tierce	1.2	Mn third	16
(3,1)	Quint	1.5	Fifth	6
(4,1)	Nominal	2.0	Octave	3
(4,1)	Deciem	2.5	Mj third	1.4
(2,2)	Undeciem	2.68	Fourth	3.6
(5,1)	Duodeciem	3.0	Twelfth	5
(6,1)	Double oct.	4.0	Octave	4.2

[a]In reality, the partials are somewhat more inharmonic.
[b]The decay times were observed in a 70 cm bell.

The pitch perceived of a bell is the strike note one octave above the lowest partial, the hum note. Its pitch is determined by the frequencies of the nominal, the duodeciem, and the double octave, which exhibit frequency ratios of 2:3:4. The ear interprets these partials as overtones to the strike note.

The tierce sounds a minor third in the timbre. Strictly speaking, it is a major sixth below the nominal. This partial appears to be very characteristic to the typical church bell timbre. It produces very special effects in carillons when dyads or chords are played. The result is quite different from that obtained with instruments producing harmonic spectra.

The minor third may not necessarily be present in all bells. As an innovation, bells with major thirds (or a minor sixth below the nominal) have been recently constructed in The Netherlands. They are supposed to sound better in carillons playing melodies in major tonality.

CHAPTER 8

Room Acoustics

I. INTRODUCTION

When most people mention acoustics they mean nothing but the acoustical properties of rooms. Similarly, when confronted with the term *music acoustics*, many people believe that the research object is merely the acoustics of music halls. By now the reader knows much better than that. Music acoustics embraces all acoustic aspects of music communication by sound signals, and consequently, also the acoustical properties of music halls represent an important part of the area of interest.

It is actually quite rare that we listen directly to the sounds generated by the music instruments. The sound reaches our eardrums only after having traveled around for awhile in the room in which the music is being played. As with other enclosed air volumes, those captured in rooms act as resonators, even though the dimensions are huge as compared with music instrument resonators.

II. REFLECTION

It was mentioned before that sound can be reflected when it hits an obstacle to its propagation. For sound reflection the same principle applies as for light reflection: The angle of incidence equals the angle of reflection (see Figure 8.1). In a room, reflection occurs when the sound collides with walls, floor, and ceiling. The reflections give rise to resonances, so it is actually such a resonator of very large dimensions that the sound has to penetrate before reaching the listener's ears. The sound transfer characteristics, i.e., the room acoustics, are therefore highly significant for the shape of the sound reaching the listener.

Reflection occurs not only because of walls, floor, and ceiling, there are often numerous other obstacles in a room as well. A condition for reflection is that the obstacle is not too small as compared with the wavelength of the sound. Thus, large obstacles are needed for reflection of low frequencies, whereas smaller obstacles reflect only higher frequencies.

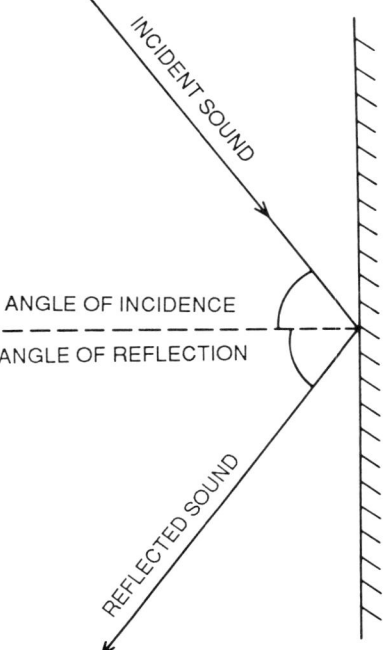

Figure 8.1 Angles of incidence and reflection.c

A. Echo and Reverberation

In large halls, special phenomena can be observed that are directly dependent on sound reflection. *Echo* means that one perceives the reflected sound as separated in time from the direct sound. The condition for perceiving two sounds as separated in time is that they are more than about 100 msec apart. Whether there is an echo therefore mainly depends on the room dimensions.

The speed of sound propagation in air is about 340 m/sec. In other words, the sound travels a distance of 1 m in 1/340 = 0.00294 sec, or approximately 3 msec, so in 100 msec it travels about 30 m. If the distance back and forth to a reflecting surface is about 30 m, i.e., if the room length is at least 15 m, an echo may occur. Therefore, it is only in a large hall that we can expect echoes. However, echoes do not necessarily occur in all large halls; it is only when there is a clear silent interval between the direct sound and a late major reflection that we hear echoes.

Sound reflection, of course, occurs also in smaller rooms. Under such conditions the reflected sound comes closer than 100 msec to the direct sound. Also, there are generally a number of reflections arriving before 100 msec that originate from various objects in large halls. In such cases one speaks of *reverberation* rather than echo.

II. Reflection

Mostly, the reflection is far from complete; rather, some sound is absorbed in the wall or in the obstacle, and some sound penetrates it. For this reason the sound loses some amplitude at each collision with a reflecting surface. In addition, the sound also loses some amplitude while traveling through air; a small portion of the sound energy is transformed into heat, which is generally referred to as *air absorption*. We will return to this phenomenon in a while.

In Chapter 2, we mentioned that the sound intensity decreases with the square of the distance, or by 6 dB per doubling of the distance. Thus, the reflected sound is reduced more and more in amplitude, the farther distance it has traveled. A long series of reflections, arriving more and more densely in time and of gradually decreasing amplitude, therefore arrives in the listener's ears when the sound source has ceased. The *reverberation time* is the time needed for the sound level to drop by 60 dB from its initial value after the sound source has ceased (see Figure 8.2). In a room, it varies more or less with frequency.

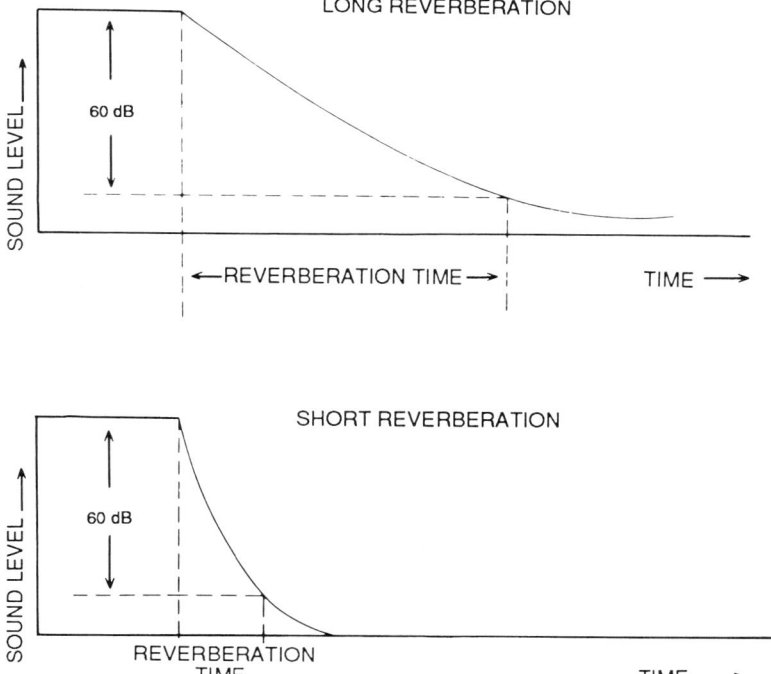

Figure 8.2 Definition of reverberation time. The curve shows how the sound level decreases after the sound had ceased. The reverberation time is the time needed for the sound level to drop by 60 dB. Normally the sound level does not decrease as evenly as in these graphs, so often one needs to draw an average curve to determine the reverberation time.

III. ABSORPTION

The reverberation time thus depends on the amount of sound energy that is lost at the reflections. A material's ability to suck sound is specified in terms of its *absorption coefficient*. It is defined as the ratio between absorbed and incident sound energy so that a low absorption coefficient implies that the material has a poor capability of absorbing sound; it throws most of it back again.

In most materials the absorption coefficient varies with frequency. This is illustrated in Figure 8.3. Soft, porous materials such as carpets and curtains absorb high frequencies whereas heavy, hard, and stiff materials reflect all frequencies. If the sound hits a wall made by a hard and heavy material, such as concrete, it is very efficiently reflected, no matter what the frequency is. In this case the reflected sound is approximately as strong as the incident sound, and it is only the travel distance and the air absorption that reduces the amplitude of the sound as it travels around between reflection surfaces. Thick and porous materials such as thick

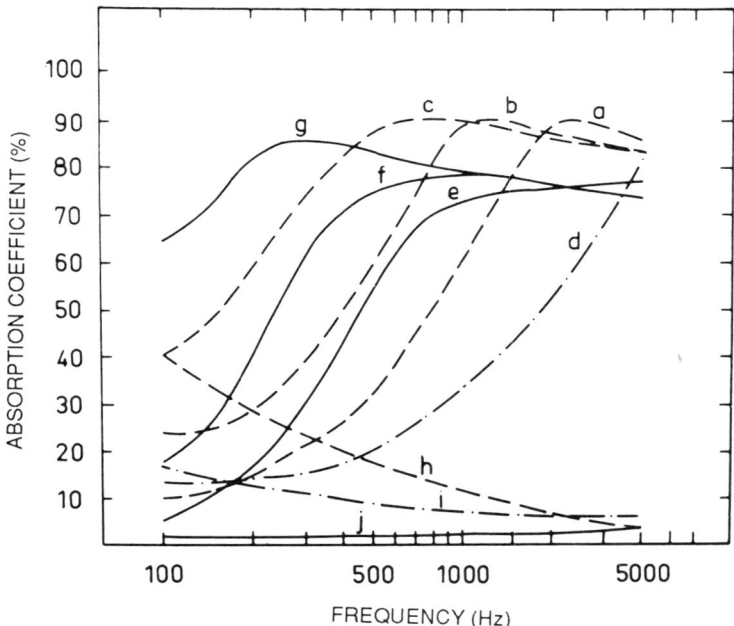

Figure 8.3 Sound absorption capability of different materials—the more porous and thicker, the more absorption also at low frequencies. The absorption coefficient, plotted along the vertical axis, indicates how large portion of a sound wave is sucked or absorbed rather than reflected, when it collides with the surface. The graph shows the absorption coefficient for different materials: (a) 2.5-cm-thick felt, (b) 5-cm-thick felt, (c) 10-cm-thick felt, (d) thick carpet, (e) 2.5-cm-thick mineral wool, (f) 5-cm-thick mineral wool, (g) 10-cm-thick mineral wool, (h) 3-mm glass sheet, (i) ordinary floor of wooden planks, (j) concrete wall. (After Björklund and Fintoft, 1966.)

III. Absorption

mineral wool have higher absorption coefficients, particularly for higher frequencies. The thicker and more porous a material is, the farther down in frequency it acts as a hungry sound absorber.

A wall covered with Dutch tiles is hard as glass, so it reflects sound efficiently at almost any frequency. If one sings in the bathroom or in a swimming hall, the entire spectrum is thrown back to one's ears. For this reason the voice of the bathroom singer sounds extremely full and resonant, particularly in his or her own ears. The opposite condition one can find in the wardrobe, where the clothes absorb most frequencies, particularly the higher ones. What one can hear of one's own voice there is far less impressive. As people seem to like hearing their own voices loudly, and decorated with many high-frequency components in the range of the singer's formant (see Chapter 3, Section III), wardrobe singers are very rare.

We mentioned air absorption earlier. It is very small but increases with frequency. At 10 kHz it amounts to 1 dB/100 m. Above 1 kHz it is affected by the ambient moisture; at 40% moisture it is about 10 times higher than at 0%, or 1 dB/10 m at 10 kHz. The air absorption contributes to the attenuation of also undesirable secondary sounds of high frequency, e.g., "scratch" in the sound of a violin or of a voice. These sounds are often easier to hear close to the source than at a distance in a concert hall.

The walls' ability to reflect sound is a factor of very great significance to the reverberation time. If much sound is lost as soon as the sound hits the ceiling, floor, or walls, the amplitude is, of course, considerably reduced after each collision. As a result, the sound dies away rather quickly when the source has ceased. And conversely, if only a very small amount of sound energy is lost in each collision, the sound retains most of its amplitude, so the reverberation time will be longer.

The last mentioned effect is often used to shorten the reverberation time in different frequency regions in a room. By putting mineral wool of appropriate thickness behind perforated plates, a portion of the incident sound is absorbed, and hence the reverberation time is shortened.

It is not only walls, floor, and ceiling that may absorb sound. Also the furniture may eat sound, particularly at higher frequencies. It is a common observation that the room acoustics, particularly in greater halls, is strongly dependent on whether an audience is present. This effect is often very striking in large stone-walled churches. The reverberation time is considerably shorter or more "dry" when such a church is fully seated than when the same hall is empty. This effect is due to the fact that all clothes are more or less porous, and therefore they absorb sounds at higher frequencies. In many concert halls with long reverberation time, one tries to reduce the difference in reverberation time between these conditions by upholstering the seats such that they absorb approximately the same amount of sound regardless if there is a person sitting on the seat or not.

Figure 8.3 shows that it is not only porous materials that absorb sound, but also materials with a dense and hard surface. The reason is that such materials can be brought to vibration. It takes a certain amount of sound energy to bring an object to vibration. Thus a sound loses energy and is attenuated when it causes an object to vibrate. Consequently, this is another case of sound absorption. Examples of surfaces that are easily vibrated by sound are windows, thin plates such as plaster walls mounted on widely spaced ribs, and wooden floors laid on widely spaced beams. It is only in cases when these surfaces are stiff, thick, and heavy that they are hard to vibrate and hence not sound absorbing.

The willingness of surfaces to vibrate is generally frequency-dependent: They act as resonators. Therefore only sounds near the resonance frequencies are absorbed, and other frequencies are reflected. As a result, the reverberation time for these resonant frequencies is shortened. Often low-frequency sounds, or bass tones, are most susceptible to this type of absorption. For this reason acousticians often call windows and thin vibrant walls *bass absorbers*. The reverberation time in a room containing many bass absorbers gets short in the bass region. This is often unfavorable. The sound quality tends to be thin or lacking strong bass components in such rooms, and this may cause problems for music. An easy method of finding out whether an object acts as a bass absorber is to strike it. The tones then resounding in the object give a hint as to what frequencies it likes to vibrate with and thus absorbs.

Air cavities behind thin walls also absorb sound. The reason is that such cavities also act as resonators. They have their own resonance frequencies at which they prefer to vibrate. We will return to this later. If sound of these frequencies reaches such a cavity, it starts vibrating at these frequencies and hence absorbs them.

A similar situation should also be mentioned. Basically it is the same phenomenon as absorption by vibrating surfaces. The sound may get stuck not only in the tiny hairy holes in a porous material. It can also get arrested in larger cavities such as vaults connecting to a greater volume, e.g., the edges of a vault above the choir gallery in a church. What happens in such cases is that too much sound is reflected by the edges and swims around within the vault. As a result, too much of the direct sound is prevented from reaching the listeners outside the vault.

IV. CALCULATING REVERBERATION TIME

What do we actually mean when we speak about the acoustics of a room? We probably mean the room's contributions to the quality of the sound that one can hear in that room: echoes, reverberation tails, etc. There is a great number of such effects, and their relevance to the sound perceived is not known in great detail yet. The most important characteristics of the acoustics of a room is related to two factors: the *reverberation time* and the *timetable of the reflections*. We will return to this timetable in a moment.

It is possible to make a fair estimate of the reverberation time, provided that two factors are known: the volume of the room and the size of the absorbing surface. The reverberation time is directly proportional to the room volume and inversely

IV. Calculating Reverberation Time

proportional to the absorbing surface. In other words, if the volume is halved, the reverberation time is also halved, and if the absorbing surface is halved, the reverberation time is doubled. In a formula called *Sabine's formula* after its inventor, the equation is:

$$t_r = C \cdot \frac{V}{S_a} \tag{8.1}$$

where t_r is the reverberation time, V is the volume in m³, S_a is the absorption surface in m², and C is a constant, which is mostly close to 0.16.

In a very large hall such as a large cathedral the reverberation time may be as long as 5 or even 10 sec. In a small church with "dry" acoustics it may be 1 or 2 sec, and in a normal living room it is generally in the vicinity of 0.5 sec.

Because the reverberation time in a room is influenced by the frequency-dependent sound absorption, it is rarely identical for all frequencies. In most rooms it is shorter at high frequencies, so one single value as those just given offers very rough information. It is more informative to specify reverberation time versus frequency in terms of a *reverberation time curve*. An example is provided by Figure 8.4.

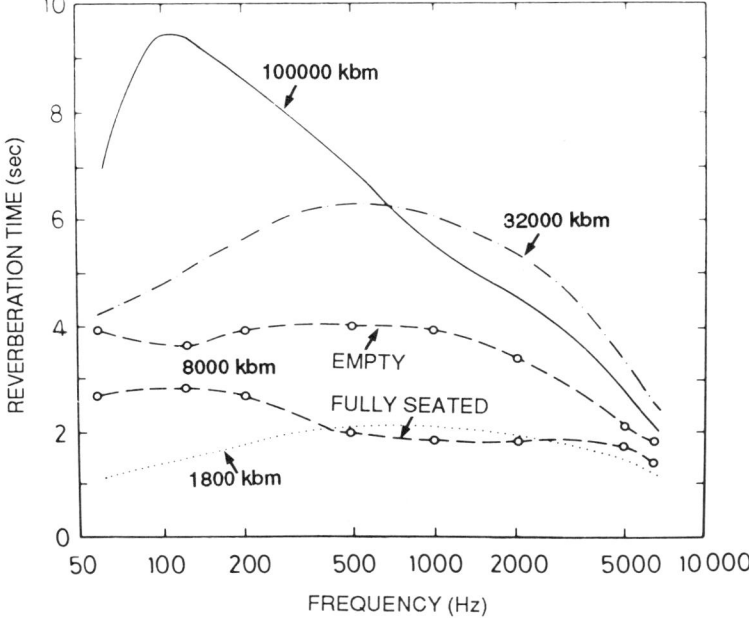

Figure 8.4 Reverberation time curves for some halls of different sizes. *Solid curve*, 100,000 m³; *dot-dashed curve*, 32,000 m³; *dotted curve*, 1,800 m³. (Data from Meyer, 1972.) The two *dashed curves* pertain to a church, 8,000 m³ measured empty and fully seated (*upper and lower curve*, respectively).

The figure illustrates the significance of the volume; large halls have long reverberations. As sound is generally much more efficiently absorbed when it hits a surface than when it travels in free air, the reverberation time should be longer in large volumes, where the distances between the limiting surfaces are long. The dependence of the sound absorbing area is also exemplified in the figure and easy to understand; if much sound is absorbed when the sound collides with a surface, the sound must be attenuated at the collisions. The effect of an audience that was mentioned before can also be observed.

A. Reverberation Radius

Most of us have probably noticed that it is easier to hear what people are saying in a room with a long reverberation time, if one is close to the talker. This is actually a consequence of the effect of the travel distance, which decreases the sound level by 6 dB per doubling (see Chapter 2).

In a reverberant room far away from a sound source, the direct sound is, of course, far weaker than the reverberating sound, but close to the source. However, the direct sound must be the strongest. At a certain distance, the direct and the reverberating sounds are equal in amplitude. This distance is called the *reverberation radius*, R_r. It depends on the reverberation time:

$$R_r = 0.056 \cdot \sqrt{\frac{V}{t_r}} \qquad (8.2)$$

where V in m³ is the volume of the room and t_r is the reverberation time. As t_r is frequency-dependent, R_r is also.

This formula is quite important, so let us get acquainted with it by using it in a concrete example. Imagine a medium-sized hall, 10×10×5 = 500 m³ and a reverberation time value of 1.5 sec for relevant frequencies. Then, the reverberation radius can be estimated as

$$R_r = 0.056 \cdot \sqrt{\frac{500}{1.5}} = 1.02 \text{ m}$$

If in one way or the other we manage to reduce the reverberation time to half, the reverberation radius increases to 1.45 m. In a hall 20×40×10 = 8,000 m³ with a reverberation time of 3 sec, the reverberation radius is 2.9 m. As we have seen, there is a relation between reverberation time and the volume of the room, and if one calculates for a little while, it becomes evident that the reverberation radius is dependent on the absorbing surface S_a only:

$$R_r = 0.14 \cdot \sqrt{S_a} \qquad (8.3)$$

If the absorbing surface is multiplied by 4 (e.g., because an audience enters the room), the reverberation radius is doubled.

The reverberation radius is very revealing as to how difficult it is to perceive speech. As long as the distance to the speaker does not exceed $3.5 \cdot R_r$, there is generally no problem in hearing what is being said, although the reverberation sound is about 10 dB stronger than the direct sound at this distance. However, at greater distances difficulties occur. Given this significance of the reverberation radius and its dependence on the absorbing area, it is obvious that a straightforward solution to speech communication problems caused by room reverberation is to increase absorbing area.

It is quite remarkable that we can perceive speech without problems even when the reverberation sound is 10 dB louder than the direct sound. This is because the direct sound has a privileged position in our auditory perception due to its early arrival. Our hearing organ seems to apply the old principle "first come, first served."

The reverberation radius also tells another interesting thing about the acoustic conditions in a room. Beyond the reverberation radius, the reverberating sound completely dominates over the direct sound. There, the sound level is completely independent of the distance to the sound source, so the distance from the sound source does not affect the sound level any more in reverberating rooms. It is only for distances shorter than one reverberation radius that the principle is valid that a doubling of the distance to the sound source reduces the sound level by 6 dB.

V. MODULATION TRANSFER

When we transfer messages coded as sound sequences, as in speech and music, the information is almost never in the stationary parts of the signal. Rather it is the changes of the signal that are revealing. A long reverberation may apparently conceal some of these changes. By combining reverberation times with hearing theory, an interesting measure has emerged called *modulation transfer index* (MTI). Modulation here stands for level variation. In an attractive way this measure takes into account how sound communication is affected by reverberation.

When we listen to sound, the sound is filtered through the critical bands of hearing, as we saw in Chapter 3. The essence of the acoustic information lies in the variations of the signal level in each of the critical bands. When this level varies in a meaningful way, we recognize sounds and understand them. In other words, it is the patterns of level variations in the critical bands that we interpret and recognize when we hear familiar sounds and understand what people are talking about. The level variations in the output signals from the critical bands are therefore the key to our capability of communicating by means of sound signals as we do in speech and music.

It is not hard to imagine what happens when a sound signal is submitted to reverberation. The level in the individual critical band will not decrease as quickly as in the original sound. Rather, the decreases will be dressed up in more or less long tails due to reverberation, as illustrated in Figure 8.5. In this way the reverberation distorts the familiar patterns, and we may have problems in recognizing them.

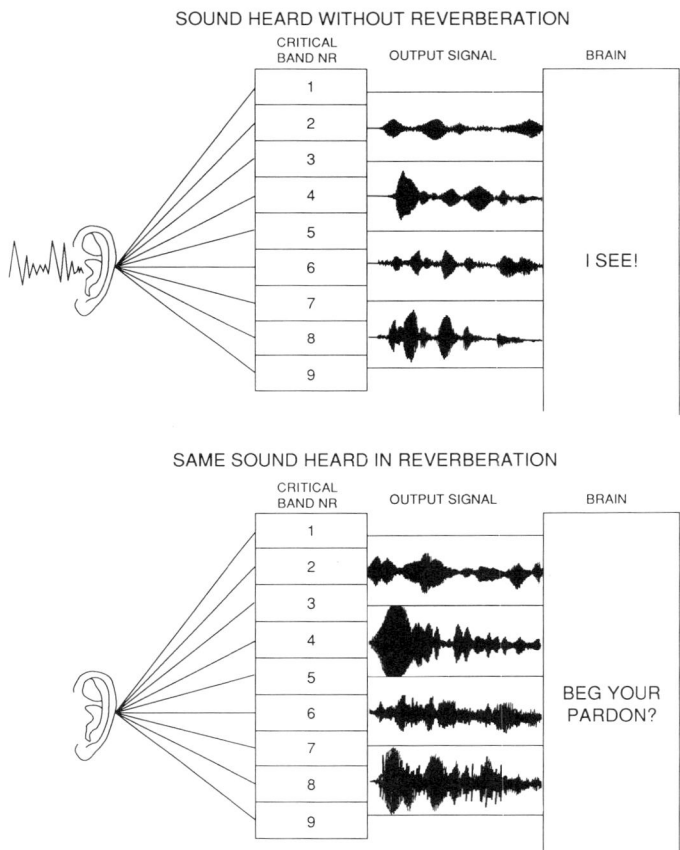

Figure 8.5 Effect of reverberation on band pass filter analysis of a speech signal simulating the critical band filtering of hearing. After filtering the input the signal is converted into amplitude modulated signals from each of the critical bands. The interpretation of the signal is based on this set of amplitude modulated patterns which are quite clear if the signal is not reverberated (*upper figure*). If the signal is reverberated (*lower figure*) the amplitude modulation patterns are destroyed in that the modulation valleys are filled out by reverberant sound. As a result, the neat amplitude patterns become more difficult to recognize and hence interpretation becomes more difficult.

One way of measuring the MTI is to emit a noise varying in sound amplitude between zero and a maximum value a certain number of times per second and record how much of this modulation survives the reverberation at a given point in the room. The MTI is a number reflecting how much of the level variations in the critical bands is preserved in a point of observation. The more reduced the level variations are in a point, the greater the difficulties to interpret correctly sound signals as speech and music.

It has been demonstrated that given the MTI, the speech intelligibility can be predicted with a reasonable accuracy. And not only that, but by means of computer programs one can also predict the MTI in rooms that only exist in terms of blueprints. It is sufficient to feed the information on the room shape and dimensions and the absorption characteristics of walls, floor, and ceiling into a computer program.

In the same way as it takes time for the sound to die in a reverberant room, it also takes time for the sound to reach full amplitude beyond the reverberation radius. We now know that the reverberation tails are a major threat to speech intelligibility. But the first sound arriving after a silent interval does not suffer from any reverberation tail. Therefore, such sounds must be more efficient in carrying an undistorted message than sounds that follow immediately after other sounds.

Some priests and some politicians are professional reverberation talkers. The former is often talking in churches with long reverberation. The latter is used to speaking in outdoor mass congregations where the public address system with long distance between many loudspeakers add reverberation tails to the sound. The typical speech behavior of these persons is acoustically quite interesting. The speech is typically divided into short sound sequences interleaved with very long silent intervals. The point would be to avoid cluttering the air with two long sound sequences that will ruin each other with their reverberation tails. Another common trick is to speak exceedingly slowly; this gives the sound some extra time to die away in the modulation valleys. A third trick is to avoid variation of fundamental frequency, which may have the effect that strong reverberation tails from very reverberant partials hide less reverberant partials.

Reverberation has also been influential in many music compositions. In music composed for long reverberation times, chord changes are often rare, so the harmonic tempo is slow. The advantage is that only one consonant chord is sounding at a time rather than a very dissonant combination of one chord plus the reverberation tails of one or more other chords.

VI. TIMETABLE FOR REFLECTIONS

Let us imagine that you clap your hands at one end of a long hall. The sound of the clap then scatters in all directions as a sphere with an increasing radius around the source. The first thing that you will hear as a listener in this hall is the *direct sound*. The first reflection is probably the one provided by the wall behind you. After some more fractions of a second, other reflections will arrive. They stem from other reflecting surfaces: floor, ceiling, walls, furniture. The first reflections arrive one by one; the later ones arrive in a continuous sequence. Figure 8.6 shows a typical example.

The timetable for the first reflections has been found to be very relevant for how the sound is perceived. The reflections arriving within the first 100 msec after the direct sound produces a clear and distinct character, and they enhance speech intelligibility. The reflections arriving later than that have the opposite effect. They make the sound unclear, jumbly, and muddy.

In 100 msec the sound travels about 34 m in air. This means that all reflecting surfaces are advantageous that offer a straight-line travel distance from the source to the listener that is no more than 34 m long. Note that sounds can have been reflected many times before they reach a listener. Therefore, it may take quite a time before the last reflection arrives. This is obviously the reason why reverberation times sometimes are very long.

It is very essential to the clarity that the direct sound leads the train of reflected sounds that reaches a listener. This condition is fulfilled for all positions from which it is possible to see the source. For this reason it is advantageous to place

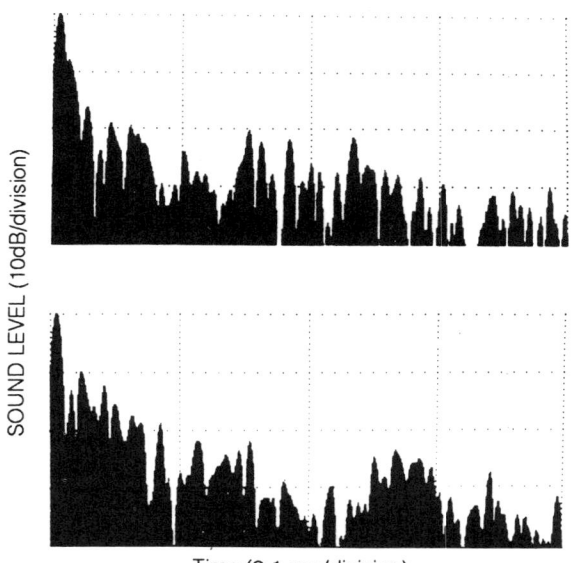

Figure 8.6 Recordings of the direct sound and the early reflections at 1 m distance in two concert halls. Each peak in the diagram refers to one individual reflection. (From Gade, 1986.)

choir and orchestra on a podium or a gallery, such that the listeners can see the musicians. This is also common practice in most music halls and churches.

Placing an organ on the floor in a church with a very high ceiling is not a good idea, because the reflections from the ceiling will probably be too late to be propitious. Placing a little ceiling over the pulpit is a good idea, because it produces nice early reflections, adding to speech intelligibility.

Cousins of the small pulpit ceilings can often be seen in concert halls where reflectors have been suspended over the podium to produce early reflections. Conversely, it is a mistake to apply sound absorbing surfaces near the sound source; the possibilities available for producing early reflections should be taken advantage of. Thus, sound-absorbing carpets and sound-absorbing ceilings or walls of a choir gallery are acoustic mistakes.

The need for early reflections should be observed when the location of a sound source is determined. Locations near a wall and not far from the ceiling are generally advantageous. However, walls are not always good sound reflectors. As we have seen, they can also be sound traps, such as in the case of vibrating windows.

Modern computer technology has brought very powerful tools to room acousticians. It is not only possible to predict the MTI by computer programs, but also the timetable for the reflections in various positions in a nonexistent, although projected hall. The input are the location, dimensions, and sound reflection characteristics of walls, floor, and ceiling. The computer program shows how the sound rays travel out from the source. Of course, the effects of different source locations, wall angles, and building materials can be tried. The program thus offers estimates of how densely in time the sound reflections arrive in different points of observation.

VII. ONE SOUND PER EAR PLEASE!

When listening to undisturbed speech, it is preferable to have the same signal in both ears (i.e., monophonic recordings). In the case of music, however, the opposite is generally preferable and likewise for speech recorded in noisy surroundings. Actually, it is an almost hopeless task to follow from a monophonic recording a speaker who was moving about; a stereophonic recording is far more helpful.

If the sound reaching the right ear is a faithful copy of that reaching the left ear, much of the information on directivity is obviously lost (see Chapter 3). The less similar the left and right ear sound signals, the more directional information and the better the intelligibility. This is no wonder. Directional hearing is very important when it comes to discerning what a person is saying who is talking in a noisy surrounding such as a merrily chatting party. Also, hearing directions must be essential for listening to music in which one instrument is playing a melody while other instruments are playing concerting melodies or interesting chord progressions. In such cases it would be bothersome if our hearing could not separate the various parts.

How can directional hearing be promoted then? Let us think of the ceiling in a concert hall; it is unlikely that it creates any marked dissimilarity between the sound reflected to the right and to the left of the room's length axis. Walls have a better potential, at least if the room is not too symmetrical and naked. Further, it should be helpful if instruments which are supposed to be distinguishable for the ear are geographically separated on the stage.

A typical expression of a listening attitude is to tilt one's head a bit to the side. This may be advantageous from an acoustic point of view, at least under certain conditions. In a symmetrical room where the walls are similar, a normal perpendicular head position will offer similar signals to both ears. If one tilts one's head, one ear will hear the sound that has been reflected by the floor and the other will catch sound which has been reflected by the sidewalls and the ceiling. These signals are likely to be different. As a matter of fact, this little trick is quite helpful in some difficult listening situations.

VIII. PODIUM ACOUSTICS

Lately, the perspective of music room acoustics has expanded somewhat. Now, questions concerning not only listeners but also performers are being taken into account. This area of room acoustics can be called *podium acoustics*.

To what degree are great sound delays acceptable on a stage? Things get tricky in ensembles that play without a conductor and where all members cannot always see all fellow instruments. It seems that in such cases the trick for keeping synchronization between instrument groups is that every player constantly listens for how the most vivid voice is being performed and adopts the timing of that voice as their own timing. Then, this voice can be more free to vary timing in accordance with musical judgment and imagination. We will return to these aspects of music performance in Chapter 10.

The demands for synchronization in ensemble playing are actually no less than enormous. A consequence of the masking effects mentioned earlier is that asynchronies as small as 20 or 30 msec can be systematically used for musical expression in ensemble performance; a soft instrument may act as a solo instrument provided that it leads by 20 or 30 msec over the accompanying instruments. The reason is that the masking threat from accompanying instruments generally appears only after their tone onsets have been completed. This is because the sound level tends to be weaker during the onset. By leading by 20 or 30 msec over a loud accompaniment, a soft instrument can still carry a solo part. The conclusion is that the control of timing must be better than 20 msec in ensemble music. This time amount appears as something of a magical limit in music making!

As sound travels through air at a rate of about 3 m/msec, it will reach about 7 m during these 20 msec. Therefore, players seated more than 7 m apart may run into synchronization problems in playing, at least if they cannot see each other. It is clearly for good reasons that orchestras pay their conductors reasonably well and call them maestro. They are badly needed not only for establishing one single musical understanding of the piece played, but their help is also required for achieving synchronization when the ensemble is so large that the 7-m boundary is exceeded for physical reasons.

IX. SELF/OTHERS BALANCE

Both for synchronization and for ensemble music performance in general, it is significant how loudly the musicians can hear themselves as compared with how loudly they hear their fellow players. This ratio has been called the *self/others balance*. Musicians hearing their own instruments too softly are prone to increase their level of playing, whereas the opposite is likely to result when they hear their fellow musicians too softly.

There are good reasons to assume that the self/others balance is also very significant to the synchronization in ensembles. For instance, it would be difficult to achieve a correct timing if the musicians cannot hear their fellow players.

The relevance of the self/others balance to tuning in choral singing is illustrated in Figure 8.7. The experiment was to sing in unison with a choir sound as soon as it appeared in the earphones. The level of this choral reference was varied, but the subjects were asked to sing at a constant sound level all the time throughout the experiment. Considerable intonation errors were caused by some self/others balance variations. The errors became disastrous when the singers could no longer hear their own voices. This result is hardly surprising. In the other extreme, the situation turned out to be less dangerous, i.e., when the sound of the fellow singers was a bit hard to perceive. In that situation musical memory and ear would provide the help needed, at least to a large extent.

How can good podium acoustics be achieved? As yet, there is no good answer. It appears clear that musicians need to hear the direct sound from each other without excessive delays. It also seems clear that early reflections, within about 100 msec, helps ensemble playing.

It appears important that all musicians are offered early reflections, regardless of where they are seated in the ensemble. Often ceiling and back and side walls offer such reflections, provided they are not too far away; according to what we saw before, the limit is about 7 m. If the ceiling is considerably higher than this, reflectors above the ensemble may be advantageous.

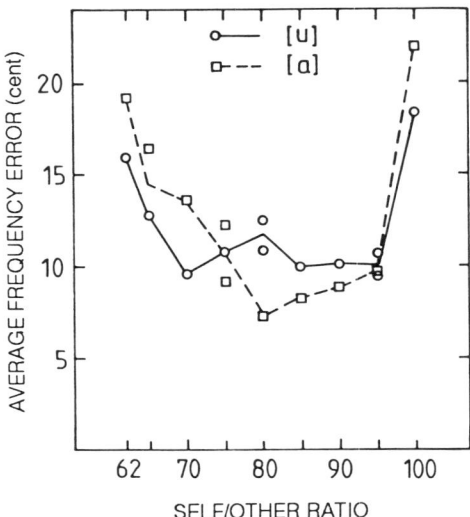

Figure 8.7 Magnitudes of fundamental frequency errors that singers made when they tried to respond in unison with a reference choral sound presented at different sound levels. Subjects always responded at a constant sound level by singing either the vowel *u*, as in who'd, or *a* as in far. Markers along the horizontal self/others balance axis represent 5 dB. Substantial errors appear when the self/others balance is unfavorable, particularly when singers can no longer hear their own voices but also when their own voices masked the reference choral sound. (From Ternström and Sundberg, 1988.)

X. MUSIC COMPOSITION AND ROOM ACOUSTICS

The acoustics of a room is determined by the construction, as we have seen. Churches have been built in different styles in all times, and it is not astonishing that also the room acoustics have shifted through the ages. This fact has had some consequences for the composition of church music.

The very long reverberation time in the old Italian Renaissance cathedrals was an acoustic reality for the church music composers of that era. Some composers certainly created their music with this type of room acoustics in the back of their heads. This type of music therefore often sounds best in churches with a long reverberation time.

The many galleries in these churches were also taken advantage of in many compositions: The choir was divided into two or more groups that concerted with each other, each from a separate gallery. Further, the chords changed relatively rarely in these compositions.

During the Baroque era, the cathedrals tended to have shorter reverberation times because of the abundance of broken contours. In church music from this time, rapid passages occur frequently and the harmonies change more often. If this type of music is performed in cathedrals with a long reverberation time, tones belonging to different harmonies merge into a dissonant, timbral disaster.

The situation becomes even worse when music composed for a string quartet is performed in a large cathedral. Traditionally, the string quarter's normal acoustical environment is a living room. If played in long reverberation, the music tends to result in a completely meaningless sequence of disorganized sounds. The new tones blend with the previous tones belonging to a different chord. Justice to this type of music is done only by halls with a rather short reverberation time, typically in the vicinity of 1 sec or less. In a similar way, other types of music need other types of acoustics.

Most church music is composed for rather long reverberation time. The secular concert halls, such as concert and opera halls, rarely reach the long reverberation times typical for cathedrals. This may cause problems for performing many types of church music in concert halls. This is a fact that should serve as a strong argument for the Church to assume responsibility for cultivating its own musical heritage!

It is sometimes argued that the skill to build good music halls was higher in the old days. As support for this idea, some excellent examples of concert halls are generally mentioned, such as that of the Musikverein in Vienna. However, the question is whether there is anything to this assertion. Probably the truth is that patience was lost with poor concert halls; either they were reconstructed or used for other purposes. A fresh example from our time is the New York Philharmonic Hall, which was submitted to a complete reconstruction, even including renaming it to the Avery Fisher Hall.

The acoustics doing justice to orchestra and choir music are rarely good for speech. Symphony orchestras need large halls capable of housing them. Such halls possess rather long reverberation times, so music written for such orchestras is generally composed for long reverberation times. Speech intelligibility, however, requires a short reverberation time. Ensemble performance raises certain demands on the self/others balance and on reasonable stage distances that are not needed for spoken performances. Music and speech thus need different types of room acoustics.

Halls are often built to serve both speech and music, thus combining the city's need for a congress hall with that of a music hall. In such cases there are often good reasons to give priority to the musical needs rather than to the speech intelligibility needs. A loudspeaker system is generally sufficient to solve speech intelligibility problems at a reasonable cost. If there are problems with music performance and/or listening, loudspeaker systems are rarely enough. In such cases it is often necessary to change the reflection characteristics of the hall, which may necessitate moving walls, changing the floor, and making other radical and hence very expensive modifications. A thorough planning by skilled professional acousticians is the way to avoid such disasters.

As different types of music need different types of acoustics, the acoustic properties of a perfect concert hall should be variable. Interesting attempts along these lines have been made. In such halls the wall elements can be rotated, turning a diffusing, an absorbing, or a reflecting surface toward the hall. A disadvantage with this solution is the associated costs.

An interesting and unorthodox attempt to improve bad room acoustics has been tried out by the Swedish instrument builder Georg Bolin. He has been experimenting with "tone walls" and "macrophones." The latter are large membranes catching a broad sound perspective of the podium, above the musicians. The former are wooden loudspeakers in which the membrane is a sound board instead of a paper funnel. The tone walls are suspended around the walls and under the ceiling and are fed with the appropriately amplified podium sound picked up by the macrophones. Sound travels much slower than electrical signals. Therefore it is very important that the sound to the tone walls is delayed to an extent such that the direct sound is always the first that reaches the listener. Otherwise the precedence effect mentioned earlier (Chapter 3) causes the illusion that the sound comes from the wall.

XI. A PERSONAL RESERVATION

Before closing this chapter it should be stressed that our understanding of the demands on room acoustics associated with performing and listening to music is incomplete. Some things are understood quite well whereas others are still something of a mystery. It is certainly possible to build halls realizing the predicted reverberation time and modulation transfer characteristics, and one can also accurately predict the timetable for early reflections. However, apart from this, other room acoustic properties may be relevant but hard to measure and predict. For these reasons there is still some uncertainty in acoustic planning of music rooms.

This uncertainty should not be taken as a reason to neglect the acoustic aspects when planning the construction or reconstruction of a music hall. Our experts certainly know very well a number of arrangements that are certain to ruin the possibilities of using a hall for music performances. Entrusting the responsibility for the acoustics of a music room to chance and lack of knowledge is unfortunately not all that rare but at the same time also overtly stupid!

XII. A FINAL REMARK

It is a very remarkable fact that so little attention is paid to room acoustic conditions when music is performed. It is not at all rare that a choir performing wonderful polyphonic music is placed far from walls and ceiling in a church.

XII. Final Remark

Orchestras may have been performing for decades of years in terrible acoustic conditions with too long or too short reverberation time and poor possibilities of hearing each other at a reasonable level. Music listeners may have been seated in places where they could not even see the performer, thus robbed of the possibility to hear direct sound. In some frequently used concert halls there are sometimes even strong echoes appearing hundreds of milliseconds after the direct sound and thus severely interfering with the musical structure.

The author has been fortunate enough to work with synthesizing musical sounds and sound sequences, thus analyzing the evasive acoustic subtleties of music performance; some of these experiences will be described in Chapter 10. An ineluctable conclusion on the way home from such synthesis sessions has been that *the acoustical shape of music performance is immensely critical.* Even the slightest little departure from the expected meaningful sound pattern is prone to catch the attention of the careful listener and so may ruin the entire aesthetic effect. It must be appreciated that

Music is the art of sound.

It would be completely unacceptable to have poor lighting conditions in an art museum. It would be out of the question for an artist to work with dusty, contaminated pigments or canvasses, or any kind of brushes depriving him or her of command over the final result. Likewise,

The acoustical shape of music is sacred.

There is no excuse for neglecting the acoustical aspects of performing music. Musicians and all of us who enjoy listening to or performing music should help other people realize this!

CHAPTER **9**

Music and Electronics

I. INTRODUCTION

In our time the development of electrotechnics has brought about immense changes in almost any field of human activities. Music is no exception. In our century the sound researcher has been given significant additions to the possibilities for solving problems. It is very thought-provoking and indeed a bit moving to see how the old classical researchers in music acoustics, such as Helmholtz and Stumpf, had to muddle and fiddle around to measure and experiment with music sounds, without our electronic means. For example, if they wanted to find out how a tone is perceived in the absence of its normal onset, the solution was to have an instrumentalist and a listening panel in separate adjacent and sound-insulated rooms, close the door, ask the player to start playing a tone, and then open the door, so that the rest of the tone can be judged. And if one wanted to measure frequency, it was necessary to blow a tuning pipe of known frequency, have it sound together with the tone to be measured, and count the beats with the turnip watch in the hand.

Thus, the technical possibilities for analyzing sound have exploded during the past decades. Without any trouble at all we can now store sound on magnetic tape, gramophone records, and computers with no or low risk of distorting it appreciably. It is easy to measure and document almost any aspects of sound. Probably all people who enjoy music listening make daily use of these facilities, and many of us might be curious to find out how they work. That part of acoustics that deals with storage, reproduction, and analysis of sound is referred to as *electroacoustics.*

Three fields of technology have developed to an indispensable position for today's music culture: the possibilities to (1) store and reproduce, (2) analyze, and (3) synthesize sound. In this chapter we will give an overview of these various possibilities. An important basis for all these possibilities is that sound can be translated to and derived from electrical signals, and first we will describe how this is possible.

A basic principle within electroacoustics is that sound can be converted to electrical signals and vice versa. A microphone converts the acoustic sound pressure variations to the corresponding electrical variations, somewhat like our ear. A loudspeaker converts electrical signals to the corresponding sound so it could, although not without considerable difficulty, be compared with our voice organ. Audio tapes and gramophone records store electrical signals such that they can be reproduced, so they have some kind of remote similarity with our capability of remembering sounds.

II. ELECTROACOUSTIC CONVERSION

A phenomenon of basic significance within electroacoustics is *electromagnetism*. This phenomenon can be observed when a coil of electrical wire is suspended in a magnetic field. If one moves the coil in the magnetic field, electrical voltage variations can be observed between the ends of the coiled wire. These variations represent an electrical image of the movements enforced on the coil. If one feeds a varying voltage to a coil attached to a paper cone, the cone will move and generate sound pressure variations. This phenomenon is used in most loudspeakers.

Thus, if a voltage variation is fed to the ends of the same coiled wire, e.g., by connecting it to a signal generator, the coil will move in synchrony with the voltage variations. This phenomenon is used in so-called *electrodynamic* microphones. If a microphone membrane is applied to the coil, the coil will move in synchrony with the sound pressure variations and will generate an electrical signal reflecting the sound pressure.

A different phenomenon that is also used for the purpose of electroacoustical conversion is a property found in certain crystals that are called *piezo-electric*. If a needle is inserted into such a crystal and a motion is enforced on this needle, electrical signals are generated in the crystal. Crystal microphones and crystal pickups in gramophones make use of this phenomenon. In the microphone application, which was quite common earlier, the membrane is fastened to the crystal. In the gramophone pickup, the needle that travels along the record groove is inserted into the crystal; the vibrations are caught by the crystal, which converts it to electrical voltage oscillations.

A third phenomenon that is also used for electroacoustic conversion can be found in electrical condensers. An electrical condenser consists of two electrically insulated plates that are charged by a positive and a negative voltage from a separate charger. When the condenser is charged, the voltage across it depends on the distance between the plates. If this distance is varied, voltage variations reflecting these distance variations are produced. If, however, the voltage across the condenser is varied, the distance between the plates tends to vary in synchrony with the voltage variations. This phenomenon is used in condenser microphones as well as in those loudspeakers that are called *electrostatic*. The *electret microphone* is a type of condenser microphone. All these devices must be provided with a basic voltage to function.

In all kinds of electroacoustical conversion, whether beginning or ending with sound, the accuracy is limited. Some distortion always occurs. There are different types of distortion.

One type of distortion is the nonlinear, also called *harmonic*. The product of this kind of distortion is similar to that produced by the nonlinearity of the ear; new tones are created at integer multiples of the frequencies of the original tones. If one records a tone of 40 Hz, this distortion will produce overtones at 2, 3, 4, etc., times 40 Hz. These harmonic overtones are often present also in the sound recorded, so this type of distortion is generally a matter of minor concern.

Somewhat more disturbing is the *difference tone* distortion. It creates subharmonics according to the same principles as the combination tones of the ear:

$$f_n = n \cdot f_{low} - (n - 1) \cdot f_{high} \quad (n = 1, 2, 3, 4, \ldots) \text{ and} \tag{9.1}$$

$$g_n = n \cdot f_{high} - (n - 1) \cdot f_{low} \quad (n = 1, 2, 3, 4, \ldots) \tag{9.2}$$

From two sine tones at 1,300 and 1,400 Hz, overtones are produced at 1,500, 1,600, 1,700, etc., Hz and at 1,200, 1,100, 1,000, etc., Hz. The electronic difference tones appear both above and below the frequencies of the original sounds. Therefore, this type of distortion creates problems in sound reproduction systems.

Intermodulation is still worse. It means that a high and a low tone interact so that combination tones occur at

$$f_n = f_{high} - (n - 1) \cdot f_{low} \quad (n = 1, 2, 3, 4, \ldots) \text{ and} \tag{9.3}$$

$$g_n = f_{high} + (n - 1) \cdot f_{low} \quad (n = 1, 2, 3, 4, \ldots) \tag{9.4}$$

Thus, if two tones of 40 Hz and 1,000 Hz are recorded, combination tones will appear at 1,040, 1,080, 1,120, etc., Hz and at 960, 920, 880, etc., Hz. This produces an unpleasant roughness in the reproduced sound as we might expect; the critical bands get crowded with interacting dissonant partials.

A. Microphones

Microphones contain a membrane that is brought to vibration by the sound pressure oscillations, as mentioned. The mechanical vibrations are converted to an electrical signal by means of some of the phenomena mentioned above; in dynamical microphones the membrane vibrations are transmitted to a coil suspended in a magnetic field, in a crystal microphone to a crystal; in a condenser or electret microphone, one of the plates is the membrane itself and the other is fixed (see Figure 9.1).

II. Electroacoustic Conversion

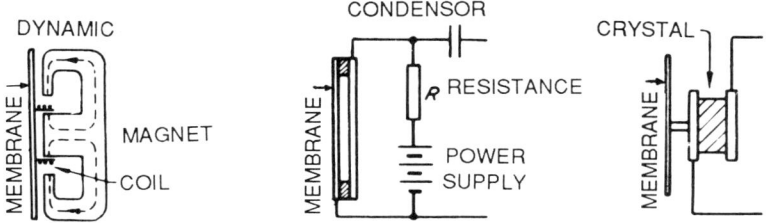

Figure 9.1   Different types of microphones.

The electric voltage variations obtained from a microphone have a very small amplitude. Therefore, an amplifier is needed before the signal can be recorded on a tape recorder or emitted from a loudspeaker. In most tape recorders, there is a separate input for microphones, which means that the amplifier needed is built into the recorder.

In some cases the microphone input cannot be used in a tape recorder or amplifier. If the input is adapted to a dynamic microphone, it would not pay to connect a crystal microphone to it. In electroacoustic terminology, a dynamic microphone has a low impedance, whereas a crystal microphone has a high impedance. High-impedance/resistance dynamic microphones also exist. In all these cases, adapters are available on the market. Very annoying difficulties and embarrassing situations can be avoided if one cements the following message in one's long-term memory:

> Before departing for recording, always set up and try the entire system of recording equipment, including every detail, even power extension cable; then pack everything and depart!

The capability of the microphone to faithfully convert sound to electrical signals is specified, among other things, in terms of its distortion. The distortion reflects the amplitude of various combination tones, so a good microphone must have a low distortion. The frequency curve of the microphone reveals how faithfully it can reproduce the amplitudes at various frequencies. A reasonably flat frequency response curve is required from a good microphone.

Microphones often have a different sensitivity to sounds coming from different directions. This directivity is specified in the *directivity patterns,* which are named according to a somewhat poetic terminology (see Figure 9.2). The directivity of the microphone is very significant to the proper choice of microphone.

Some microphones, the directional ones, are mainly sensitive to sound coming from the front. Some microphones are almost stone deaf to all sounds except those coming exactly from the front. Such microphones hear only the sounds they are aimed at, and this may be a very great help when recording in noisy rooms. Less

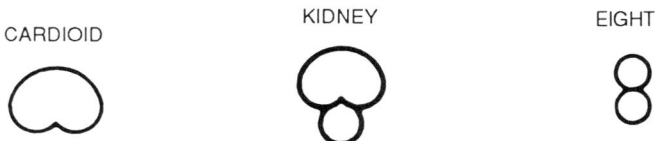

Figure 9.2 Different types of directivity patterns for microphones.

extremely directional microphones may be advantageous to use for recordings in highly reverberant rooms such as cathedrals. Some microphones have a more spherical sensitivity, and they are appropriate when sounds should be recorded that arrive from many directions.

B. Loudspeakers

How sound is produced by the movement of a loudspeaker membrane was described in detail in Chapter 2. Most loudspeakers are dynamic, i.e., a paper cone is brought to vibration because it is fastened to a coil suspended in a magnetic field, as illustrated in Figure 9.3. The electrical signal is fed to the coil, which thereby is brought to vibrate back and forth in the magnetic field. In electrostatic loudspeakers, one of the condenser plates is fixed while the other is brought to vibration by varying the voltage between the plates.

The loudspeaker is generally mounted in an enclosure. The task of this enclosure is to insulate the front side from the back side. In this way the air cannot simply slip around the loudspeaker edge so that two sound waves in counterphase are added and cancel each other. This case occurs when the wavelength is long as compared with the dimensions of the loudspeaker. Therefore, the enclosure helps the reproduction of low notes in a way similar to kettle drum and drum.

Some more improvement of the reproduction of low tones can be obtained if the cavity behind the speaker membrane is used as a resonator for lower frequencies. The box is coated with damping material so that sounds near the resonance frequency of the cavity will not be overly enhanced. In the so-called bass reflex loudspeakers, there is a hole in the resonance box. The airplug in this hole resonates with the volume in the enclosure to form a big Helmholtz resonator.

The suspension of an ideal loudspeaker membrane is infinitely flexible, infinitely stiff, and weighs nothing. Obviously none of these conditions can be fulfilled in reality. This causes some distortion and unevenness in the frequency curve. To obtain an efficient sound reproduction, the membrane must also be large as compared with the wavelength of the sound. However, it must be small as compared with the wavelength to produce an even radiation in all directions. The poor reproduction of the bass in small loudspeakers is due to the fact that the membrane is small.

II. Electroacoustic Conversion

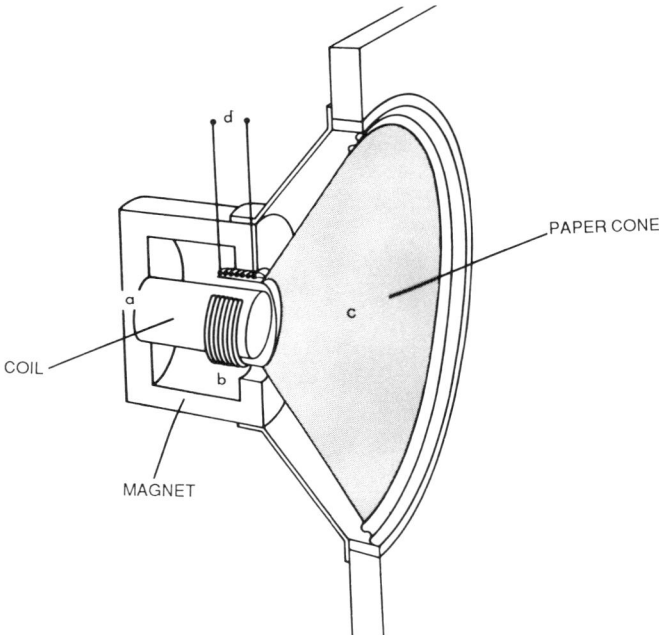

Figure 9.3 Loudspeaker. On the coil (a) that is suspended in the magnetic field (b), the paper cone (c) is fastened. It moves when an electrical signal is fed to the ends (d) of the coil. (From Hadding and Petersson, 1972.)

The art of constructing a good loudspeaker is complicated, also because of difficulties to establish objective criteria for what quality actually should mean. This would be a reason why specialists in this field rarely agree but rather tend to revert to a firm and rigid conviction, not rarely supported by strong emotions.

Loudspeakers cannot be connected to any output of an amplifier. Outputs devised for loudspeakers are generally marked with a loudspeaker sign (⊏◁) and/or a resistance of a few ohms. Dynamic earphones should be connected to an output marked "Phones."

Generally the sound reproduction quality obtained with good earphones is better than that obtained with loudspeakers, at least in stereophonic systems. In listening tests, earphones also have the advantage that it is possible to know exactly what is being fed to the ears of the listeners. This is not true for loudspeaker listening in normal rooms, where standing waves and other room acoustic phenomena contribute to the signal reaching the listeners' eardrums.

Caution is recommended when using earphones. It easily happens that very loud click sounds are produced when amplifiers and other pieces of equipment are

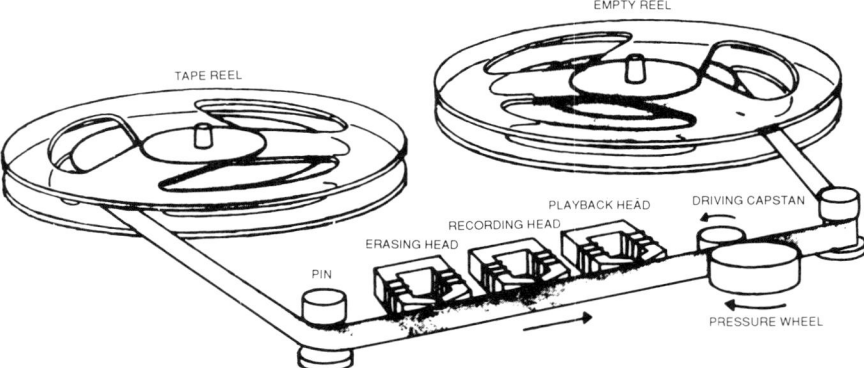

Figure 9.4 Tape recorder. (From Stensson.)

switched on and off, so the volume control must then be turned down. Otherwise both your ears and your earphones may be damaged. Also, loudspeakers can be damaged by stray click sounds.

C. Tape Recorders

The tape recorder stores the electrical signal corresponding to the acoustic signal. This is realized by converting the acoustic signal to a magnetic signal, which is fed to the tape.

One side of the tape is coated with a thin layer of ferric oxide. This side is mostly brown as rust and less shiny than the other side. The coated side is magnetized by pulling it past a magnet, the *record* head, the magnetization strength of which reflects the electric signal from, for example, a microphone. The signal is then stored in the ferric oxide in terms of a pattern of variable magnetization. The tape passes also two other heads. One is the *playback* head, which senses the tape magnetization and converts to the corresponding electrical signal. By passing the *erase head,* a magnetization can be washed out. This head is actually another record head that records a strong high-frequency tone, well above the audible range. The order of the three heads is the same in all tape recorders; the tape first passes the erase head, then the record head, and finally the playback head, as illustrated in Figure 9.4. In some less ambitious recorders, the record head is also used for reproduction, so then there are only two heads.

In case the record and playback heads are separate, it is possible to control the quality of a recording while it is being done. By means of a switch, often marked "Input/Tape," the listening unit can be connected to either the input signal or the

II. Electroacoustic Conversion

reproduction head. By switching between "Input" and "Tape," the quality before and after recording can be easily compared. Thereby, the recorded signal is heard with a slight time delay, because the tape passes the reproduction head a bit later than the recording head.

The time delay between the record and playback heads offers an opportunity to add a (highly) artificial reverberation to a recording; one records the reproduced signal from one track on the other track and listens to both tracks monophonically. By reducing the amplitude of the delayed track appropriately, some of the unnaturalness of this reverberation can be reduced somewhat.

An input switch connects one of the different input contacts, mostly marked "Microphone," "Radio," and "Aux," to the recording head. It is, of course, important to connect the proper input line to the recording head.

It is also important not to overload the recording by amplifying the input signal too much. This amplification is regulated by an input volume control. Distortion adds an unpleasant quality to the recording, which sounds "impure" or "clashing." In most tape recorders there is an indicator showing the result of the amplification in terms of a needle deflection. The needle is not allowed to pass an upper limit marked "0 dB" or by red color on the scale. Sound with a very high acoustic level may appear quite unexpectedly, e.g., when a soprano sings at high pitches or from bass instruments.

In addition to or instead of an input volume control, some recorders have an automatic input gain control marked "AGC" or "Limiter," which senses the input level and adjusts the amplification accordingly. This device is useful for recording signals with reasonably constant level such as speech but rarely for music.

The dynamic range of a tape recorder is limited by two factors. One is the background noise, which can never be avoided except in digital recorders, as we will see soon. Too soft signals are drowned in this noise. The other factor is the highest acceptable signal amplitude that can be recorded without distortion. The maximum dynamic range of a good conventional recorder is often no more than 50 dB, which is much less than what typically occurs in music.

Exaggerated fear of overloading leads to another disadvantage. The background noise becomes loud as compared with the softest parts of the music recorded. The trick is to find an optimum. For several reasons it is recommended to first make a test recording, listen carefully to the result with excellent listening equipment, and then make the real recording.

If both microphone and loudspeaker are connected at the same time to a tape recorder or an amplifier, it is important to see that the listening level is not too high. Otherwise a very loud scream is produced when the amplifier catches and raises the sound level of the loudspeaker, which then, in turn, is picked up by the microphone, and so on.

Some of the magnetization on the ferric oxide layer on the tape may scatter to adjacent tape turns on the reel, particularly if the tape is not played for a long time.

These tape echoes may be quite irritating, particularly when they arrive in a soft part of the recording just before a sudden, loud sound. The echoes of the coming loud sound may then be heard softly several times before the real sound appears. Such effects were not infrequently heard on the radio in old tape recordings of theater dramas.

It is, of course, possible to magnetize a tape by other methods than that represented by the recording head, e.g., an object made of magnetic iron. This may sometimes lead to unpleasant consequences. If one brings a tape too close to a strong magnet, disturbing sounds will appear in the recording. Many objects may quite unexpectedly turn out to be magnetic: scissors, screwdrivers, razor blades. In buses, subways, planes, and trains, very strong magnetic fields exist. A safe way to avoid ruining a tape recording is to transport tape reels wrapped in aluminum foil or in a metal cassette.

Normal tape recorders change the phase relations between the partials in a spectrum. Therefore, phase relations cannot be analyzed via normal tape recordings. For such purposes a special type of tape recorder is needed, called *FM* tape recorders (FM stands for frequency modulation).

In most reel-to-reel tape recorders, i.e., recorders where the tape is stored on reels rather than in cassettes, one can choose between different tape speeds. The main principle is the higher quality needed, the higher the tape speed. The top speed in most recorders is 15 ft/sec (38 cm/sec). This speed is mostly chosen for original recordings or for recordings intended for very accurate spectrum analysis. Normally, 19 cm/sec (7.5 ft/sec) is enough. This speed is generally the highest speed available on good amateur recorders. Lower speeds can be used for recording speech, when the speech rather than the voice characteristics is of primary concern. If one uses a different speed for the playback than was used for the recording, the result is not only a transposition in frequency, but also a less perfect frequency curve plus a reduced dynamical range.

Reel-to-reel recorders are rare these days. Instead cassette recorders are used, where the tape speed is quite low and cannot be changed. Still, the quality of the sound reproduction may be quite high, provided good amplifiers and loudspeakers are used. Much of this effect is due to the perfection of the technology.

Particularly in cassette recordings, noise reduction systems such as DOLBY, DBX, etc., are of great significance. Some of these systems compress the dynamic variations in various frequency ranges during the recording and expand it during playback. Others just cut out the high-frequency range when there is no signal there. It is important to use the same noise reduction system for both recording and listening.

Nothing is perfect. This also applies to tape recorders. One problem is always the background noise. Particularly when copies are made of copies of copies of copies, etc., background noise causes problems. Another problem is that the tape rarely has an exactly constant speed. The result is a corresponding frequency variation, called *wow*. This type of problem is most easily heard in recordings of instruments with very long and constant tones, such as slow piano or organ music. The wow is much less in expensive recorders and at higher tape speeds.

Another common problem in cheap tape recorders is that the signal suddenly just disappears and then immediately returns, so-called drop-outs. This problem is caused by poor tape feeding, resulting in a poor contact between the heads and the tape. This problem is sometimes caused also by the tape. If one suspects that a tape recorder has a poor tape transport system, it is wise to check that the problem does not come from a bad tape.

Finally, it should also be mentioned that tape recorders can be trimmed to optimally fit certain types of tape. In such cases, shifting to a more expensive tape quality may easily result in a lowering of the quality of reproduction!

D. Phonograph

If there were any space for an eighth miracle, the phonograph record would be a strong candidate. In it, the microscopic sound pressure variations within a huge frequency range are transferred to little bends in a groove! In the case of a sine tone, the groove goes a little to the left and then a little to the right, such that the needle tracking the groove will move according to a sine curve. In the case of complex tones such as in music, the groove's deviations become correspondingly complex. The overall idea is to transfer the microphone vibrations—via an electrical system—to a groove in a record, which are then tracked by the needle. The vibrations of this needle are then transformed to electric voltage variations by means of a piezo-electric or dynamic system.

There are several marvels associated with the phonograph record. One is that its frequency and amplitude range are sufficient for reproduction of music. It is no less than astounding that it is possible to mass fabricate a groove that turns right and left 15,000 times per second. We might imagine that this will be more easily achieved with a quickly rotating record, rather than a slow one. Still, it is enough with 33 rotations per minute. Thereby it should be realized that the groove speed is considerably greater near the outer periphery, where the circumference of a groove turn is long, than near the center, where this circumference is considerably shorter. For this reason the quality of a phonograph recording often drops slightly toward the center of the record.

Another astounding fact is that the record can be copied with so high accuracy. The slightest microscopic deviation from the intended groove shape causes distortion, and still, distortion is rather moderate in phonograph records. Actually, distortion will result even if there is only a tiny little grain of dust lying in the groove, which disturbs the needle vibrations. One can imagine that it must be a sin of death in music acoustics to grasp a record by putting one's fingers across the grooves.

E. Digital Recording

The weakness of a phonograph record is that a low minimum background noise level cannot be avoided. This imposes a limitation on the dynamic range. The digital recording technique has solved this problem.

Digital sound reproduction technique means that one no longer stores an electrical equivalent of the sound pressure variations. Instead, the instantaneous sound pressure value is repeatedly measured at very short time intervals, as illustrated in Figure 9.5. This technique is called *sampling*, and the number of readings made per second is called *sampling frequency*. A low sampling frequency results in a poor reproduction of the finer waveform details, i.e., higher frequencies and softer sounds. In digital tape recorders and in compact disk (CD) records, these instant sound pressure readings rather than the waveform itself are stored on memory devices in the form of numbers. Thereby, the numbers are translated to binary form, which means that every number is represented by a combination of zeroes and ones.

This means that the original sound pressure signal has been transformed into a new shape, namely, a long table of numbers. As numbers are also called digits, this type of representation is called *digital*. In an apparent attempt to keep the unknowledgeable mass out in the dark, the original form is called *analog*. This most remarkable choice of terms suggests that the real thing is not at all real, just something similar to the real thing! Those thirsting for knowledge and understanding have to accept this terminological whimsicalness as a sad fact. The device

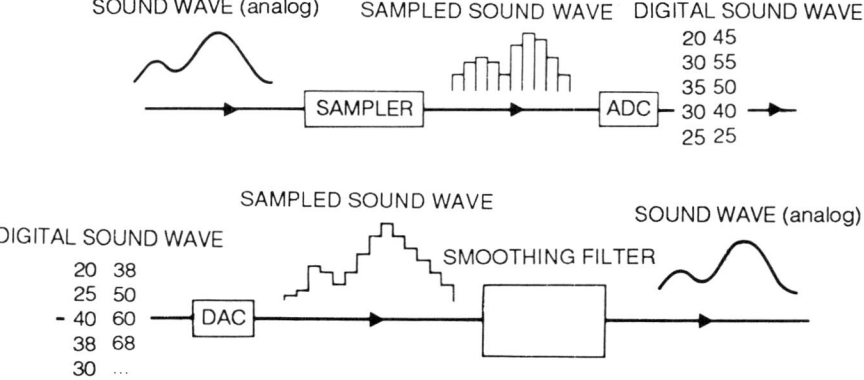

Figure 9.5 Principle of digital sampling of a signal. The sampling implies that the instantaneous value of the sound pressure is repeatedly measured at very short time intervals, the inverse of the sampling frequency. These values are stored in digital form. For reproduction, the sampled signal must be filtered so that corners of the wave form are eliminated. (From Rossing, 1982.)

translating the signal from the original (analog) form to the digital form is called analog-to-digital converter (ADC).

When reading the memory, the stored digital values are retrieved. They are converted into the original (analog) sound pressure signal in a digital-to-analog converter (DAC). This device also smoothes the wave form by means of a low-pass filtering process, which we will describe later.

An advantage with digital storage of sound is that there is no other distortion of the signal than that associated with the sampling frequency. It is enough to set this frequency to twice the highest frequency that one wants to reproduce. There is no background noise whatsoever to worry about, so the dynamic range grows almost beyond reasonable limits; instead of the impressive 50 dB of conventional tape recorders, the digital devices offer something near 95 dB! This means that when recording a piece of music there is no need for attenuating the louder parts and amplifying the softer parts to adapt it to the dynamical range of the tape recorder. Further there is, of course, not the slightest trace of wow.

The digital technique has its limitations, too. For signals with very low amplitude, the digital picture is not all that faithful. This means that sounds lying, say, 80 dB below the loudest are distorted. Learned people dispute as to the perceptual relevance of this limitation for reproducing room acoustic information, because the final tail of reverberating sound may be significant for this. One could argue that this problem should be infinitely worse in a recording of traditional type, but, on the other hand, in such recordings there is always the background noise, which kindly masks the small flaws in the reproduction of very low sound levels.

Summarizing, the digital technique has contributed an extremely valuable optional method to sound reproduction technology. It will be exciting to see which storage technology will win in the future, the CD record, in which the digital information is stored with an optical method, or the digital cassette recorder, in which the storage is electromagnetic and happens as in conventional audio tapes.

F. Dummy Head Stereophony

No sound reproducing technology can recreate exactly the sound that reaches our ears at the concert. There are several reasons for this. One is that the directivity of the ears is not the same as that of the microphone. Another reason is that loudspeakers reproduce music with the reverberation in both the recording room and the listening room. Moreover, no loudspeakers have a perfectly flat frequency-response curve. Also, sound leaks from the left speaker to the right ear and vice versa in a stereophonic recording, in which one channel is reproduced by the left loudspeaker and the other by the right. The net result of all these complications is that it is impossible to achieve a perfect match between original and reproduced sound.

One way of circumventing several of these complications is to mount two microphones in the ear canals of a dummy head. This is the idea in the so-called dummy head stereophony. It has several advantages. The directivity of the microphones is almost exactly identical with that of the ear. Also, the diffraction of the sound around the head is realistic, which is very significant for the directional hearing, as we have seen. If the recorded sound is reproduced over head phones, the problem with the double room reverberation is eliminated.

For these reasons dummy head stereophony recordings offer a much more faithful reproduction than ordinary microphones and loudspeakers. Still, not even in this case is the fidelity infinite. When listening over earphones, a small part of the outer ear and the ear canal are involved twice in the process: in the dummy head and in the listener's head. It would be possible to avoid this by placing the microphones at the entrance of the ear canal, but then the directional hearing suffers a little, because the ear canal is significant for this effect.

Apart from this limitation, dummy head stereophony recordings also have some other disadvantages. It is a bit inconvenient to wear earphones at length, and it is not all that tempting to wrap one's friends into earphones as soon as one wants to listen to music together.

The sad end of the story therefore is that it is impossible to reproduce exactly the sound that one would have heard when the music was played and recorded. Thus, the only way is to weigh one disadvantage against another, so ultimately it comes to subjective preferences. Exactly this dilemma, that, objectively speaking, there is no best solution, makes it difficult to arrive at an unanimous view of what is good and what is bad in sound reproduction. In such situations it is very typical that one develops one's own religion and that adrenaline rather than facts dominates the discussion.

III. A QUICK LOOK AT THE WORLD OF SYNTHESIZERS

Recent development in the area of electronic engineering allows the construction of exceedingly small amplifiers, filters, generators, etc. This has opened new and very significant possibilities in music culture. Among these, the synthesizer is an example that is interesting from a music acoustics point of view. Development is quick in this area so what is written now will soon be obsolete. Still, a short overview will be attempted.

A major characteristic feature of a synthesizer is that the instrument generates directly neither mechanical nor acoustical vibrations. As we have seen, all traditional instruments generate such vibrations in their strings or air columns. Instead, synthesizers generate an *electrical* signal, which is then converted to acoustic vibrations in amplifier plus loudspeaker. For the generation of the primary electric signal, different methods are used, and these characterize the synthesizer.

III. A Quick Look in the World of Synthesizers

The most straightforward method is similar to sauce cooking: one takes a tone, adds another one, adds a third one, etc., until one has obtained a timbre that meets one's needs and pleases one's taste. This principle is called *additive synthesis*. The oldest representative of this type of synthesizer is the Hammond organ, where the amplitude of each of the lower harmonics could be adjusted by sliders. The principle of additive synthesis is used also in some modern synthesizers.

The opposite of the additive synthesis is the *subtractive synthesis,* which seems very common today. It builds on the filtering principle, so the raw material is a complex tone with a great number of strong partials; the amplitudes of the partials in any frequency region of this spectrum can be modified by filtering.

It is very easy to carefully elaborate a spectrum with a given fundamental frequency on a synthesizer and then have it transposed to any other fundamental frequency represented on the keyboard. A silent assumption then is that a constant timbre will result from a constant spectrum. After having read this book this far, we know better than that. Our hearing is prone to note the frequency location of spectrum peaks. If they remain at approximately the same frequencies regardless of the fundamental frequency the timbre is perceived as constant. In other words, *frequency* rather than partial number is relevant to our perception. Therefore, the "same spectrum" presented at different fundamental frequencies does not necessarily give the same or even a similar timbre.

In some cases we have learned to disregard the fact that the spectrum changes dramatically with pitch, so that we still think that the timbre stays the same. For instance, we may think that a flute has the same timbre throughout its range even though there are huge spectrum differences between the high and low tones, as we saw in Chapter 5. Also, to some extent we can listen to a spectrum in a more absolute way. The typical feature of a clarinet tone in the lower register is that the lower even-numbered partials are weak.

A great and significant innovation of the synthesizer market was the so-called *FM technique.* Its inventor is an American composer originally trained in traditional composition, John Chowning. His idea is simple: The frequency of a sine tone is varied according to another sine wave, similar to what happens in a vibrato. However, the frequency variation is much quicker than in the vibrato.

The technical solution is to feed a sine wave signal to the frequency control input of a voltage-controlled sine wave generator. Thus, the former sine wave signal, called the *modulation signal*, modulates the latter, which is called the *carrier.* What then happens is that an entire spectrum is generated around the carrier frequency, and the frequency distance between neighboring partials equals the frequency of the modulation signal, i.e., the modulation frequency (see Figure 9.6). The phenomenon is thus similar to intermodulation distortion.

Given the principles just mentioned, we realize that a harmonic spectrum will be produced if the ratio between the carrier and the modulation tone frequencies can be expressed by small integers. Further, the more the amplitude of the modula-

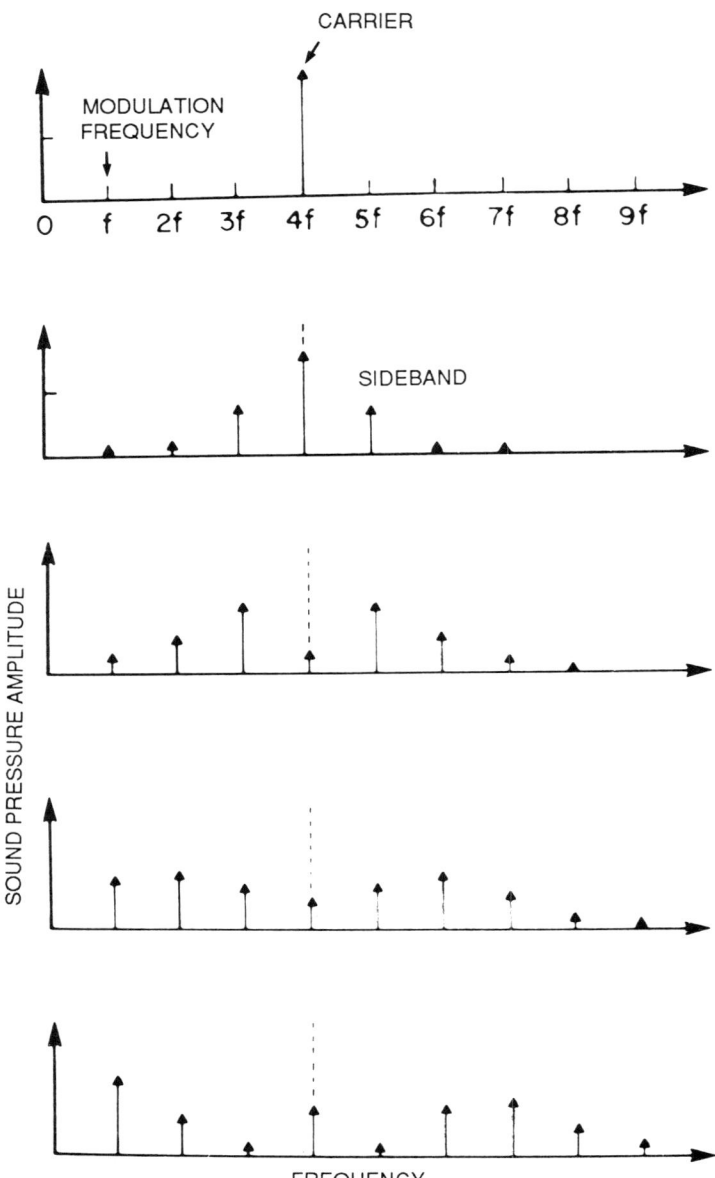

Figure 9.6 Principle for frequency modulation. The modulation, which increases from top to bottom in the figure, gives rise to new partials, so-called sidebands, which appear above and below the carrier. In this way an entire spectrum can be generated using two sine wave generators only. (From Chowning, 1977.)

tion tone is increased, the higher the number and amplitudes of the partials. If the partials land on negative frequencies, they do not disappear, but are merely mirrored at zero frequency; they appear at positive frequencies all right and the only trace of their negativity is that their phase is shifted. The FM technique is used in many synthesizers, among them several of the trademark Yamaha.

The *sampler* is another interesting variant of the synthesizer. The idea is that one records one sound per key, and this recorded sound is played as soon as the key is pressed. One can also record only one or two tones per octave and then transpose the sound to neighboring tones so that one recorded sound is used for several different fundamental frequencies. One simply reads the stored sound at different speeds, depending on the fundamental frequency. In this way the spectrum is transposed so that it is available with different pitches.

If one uses the same stored signal for different fundamental frequencies by varying the playback speed, not only the fundamental but also the entire spectrum is transposed, including all its spectrum peaks and valleys, if any. In the human voice, a formant frequency cannot be transposed by even 1% without causing a noticeable effect on the voice and vowel timbres. A semitone equals a fundamental frequency rise of about 6%, which thus is many times too much to preserve any voice timbre. For this reason samplers are hard to use for sounds characterized by formant-like spectrum peaks. The timbre will not stay the same for neighboring notes. The solution in such cases is to record a sample for each key, which, however, is slightly alien to the idea of the sampler. For many other instruments, there are no such problems. For instance, piano, bowed instruments and many wind instruments are easy to imitate using samplers.

An immense advantage with samplers is that, almost without exceptions, they can be controlled in a standardized form via a standard code and contact. This code is called MIDI (music instruments digital interface). One can therefore easily connect different synthesizers with different control organs such as keyboards, pedals, etc., and also connect them with each other.

Among the control tools, there is rich choice available. The keyboard is the most common. Some keyboards can sense how hard the key is struck so that the sound level increases if one strikes the key harder. Some keyboards also sense the pressure exerted by the finger on the key after it has reached the bottom of its motion, so-called aftertouch. This allows the player to affect the tone while it is sounding. There are also other control facilities such as pedals or mouthpieces for blowing. These tools can be used for controlling various aspects of the tone. Complemented by these means, the synthesizer becomes a very flexible instrument.

Clearly, the synthesizer is now one of the most common music instruments; it assumed the role that the piano was playing around the turn of the century and the guitar in the 1970s. But is the synthesizer really a "genuine" music instrument? If we allow time to think about it thoroughly, the answer to this question must probably be in the affirmative. One must admit that everything that is used for producing sounds in a music performance must be accepted as a music instrument, period! If the composer uses the sound of a stone falling on a wooden floor in a piece, the stone and the floor are obviously an instrument.

Sometimes one can hear people argue that the synthesizer is such a simple instrument, because there is no way of affecting the tone color. That argument is untenable, as we have seen. Also, music works successfully with instruments that differ dramatically in this regard. On the one hand, there is the human voice, the most flexible of all instruments: loudness, pitch, and timbre can all be bent *continuously* and in a way that the performer can control. On the other hand, there is the completely rigid organ, on which the player can control only one single thing while playing, namely, whether the tone should sound; all other aspects of the tone can only be changed in steps in many of the best Baroque organs. Still, the organ is rightly a highly appreciated instrument, even by the author. Between these extremes in flexibility we find the other instruments: the bowed instruments, in which pitch and loudness can be varied continuously while the resonance box remains constant; the wind instruments, in which pitch and timbre are flexible within certain limits only; the piano, in which the timbre seems to be somewhat tied to loudness. Along this scale from rigid to flexible the synthesizer is by no means extreme. Rather it is located near the middle. It can be a much more flexible and obedient instrument than the piano with respect to reacting on the player's gestures.

One could also argue that the synthesizer is an alien because the sound generation is electronic rather than natural, which would mean mechanical or acoustical, as in the traditional instruments. However, it is also hard to maintain this view after some moments' reflection. We may accept that blowing into a tube or hitting a membrane is close to nature. But how natural may it be to pull rosined horsehair over a string strung over a resonance box? Also, is electricity unnatural, despite thunder and lightning?

A certain number of properties, however, make the synthesizer a special animal in the world of music instruments. One is that it has rarely had the courage to acknowledge its own autonomous identity. It seems applicable to personalities and music instruments alike that people tend to disregard those who lack the courage of being themselves and instead resort to imitating others. As long as the synthesizer is used only to "replace" other existing instruments, it would certainly be regarded as a substitute of inferior value than the "real" instruments. However, it should have good possibilities to be a very useful instrument of its own, so it does not really need to imitate "real" instruments. This fate of being a substitute for real instruments is just glued on to the idea of the synthesizer; it is not an inherent part of it.

Another characteristic that makes the synthesizer a special animal in the world of music instruments is its sound radiation. Traditional instruments have very complicated sound radiation devices: one or two sound holes, a sound board, a resonance box, etc. The synthesizer has one or more loudspeakers. It is possible that this difference significantly contributes to the properties of the synthesizer that are relevant to its characteristics as a music instrument.

A third unique thing about the synthesizer is that, contrary to all traditional instruments, it does not contain any resonator. This may have an audible effect on the tone characteristics, particularly in the onset and decay. It may also be relevant from other points of view. For instance, the resonance frequencies are fixed in many instruments and so add spectrum peaks that are independent of the fundamental frequency. It would not be impossible that our hearing keeps track of this.

What was the answer that emerged from this discussion, then? Well, there seems to be no firm reasons to despise the synthesizer as a music instrument, particularly if it is not used merely to replace original traditional instruments. Perhaps there are also particular properties related to resonators and sound radiation that place the synthesizer in a special group of sound generators. But also as a member of that group it should have good potentials of adding new and valuable possibilities to music culture.

IV. ACOUSTIC MEASUREMENT EQUIPMENT

Physically, sounding music corresponds mainly to variations of frequency and amplitude with time. An analysis curve on a graph offers information regarding two dimensions only. Either we can see how frequency varies with time (dimensions: time and frequency) or how the amplitude varies with time (dimensions: time and amplitude). However, as we will see, there are also devices that can show all three dimensions simultaneously.

Amplitude events can be recorded by means of (sound) level recorders and oscillographs. Level recorders shows the sound level of the input signal as a function of time on a paper strip. An example is shown in Figure 9.7a. Thus, it does not display the waveform. That possibility is offered by an oscillograph (e.g., the mingograph). A common version is the ink jet writer, which throws an ink jet through a narrow nozzle, the direction of which reflects the sound pressure. An example of a mingograph recording is shown in Figure 9.7b. Normally, oscillographs have the limitation that they cannot record higher frequencies than about 800 Hz.

A recording of waveforms also containing high frequencies can be obtained from an oscilloscope, which shows the waveform as a pattern on a screen. In some oscilloscopes provided with a memory, one can connect a pen writer to obtain the waveform on paper. The oscilloscope thus does not have the frequency limitation of the ink jet writer, but detailed pictures can be obtained of only a few waves. Digital memory oscilloscopes can mostly store only rather short pieces of sound

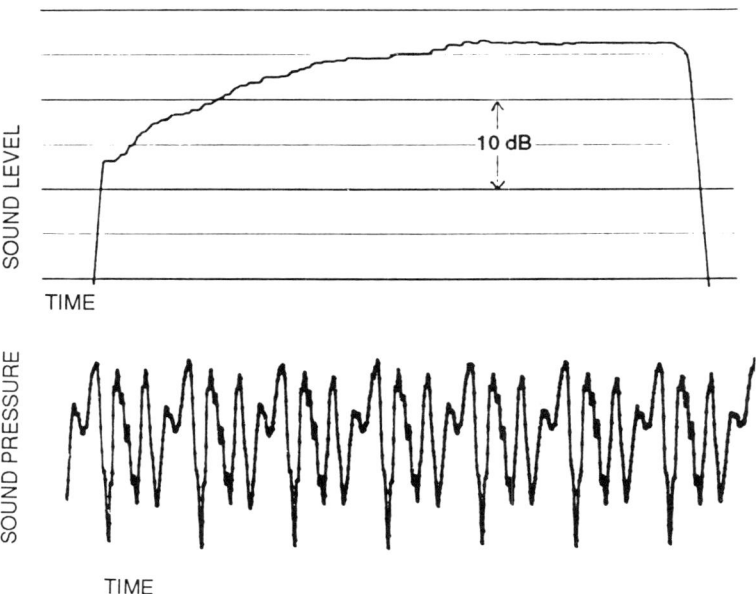

Figure 9.7 (a) Sound level recorder registration of a crescendo as performed by a singer. (b) Oscillograph recording of the waveform of a sung vowel.

and display it in its entirety or a part of it. Much of what we now know about the details of, for example, pianos has been revealed because of the availability of memory oscilloscopes. They are generally standard in acoustic laboratories.

Melody writers such as the melograph belong to standard equipment in acoustic speech laboratories. In some places there are versions particularly suitable for music. Figure 9.8 shows an example of a recording from such a device, which offers a recording of the sound level as well. Melographs extract the fundamental frequency of the signal and shows how it varies with time. Normally they work only for one-voice music, and there are often limitations as to the possible fundamental frequency range.

Fundamental frequency extraction from complex sounds is a classical problem in music and speech research. The possibilities of success are much better if the recording fulfills certain demands. In many cases, including the human voice, it saves lots of trouble and time to record the signal by means of a vibration-sensing microphone, so-called accelerometers, fastened to the neck wall below the larynx. An alternative is the electroglottograph, which records the vocal fold contact area; when the vocal folds collide, there is a strong peak in the signal, and this helps fundamental frequency extraction considerably. For bowed instruments it is advantageous to fasten an accelerometer to the bridge.

IV. Acoustic Measurement Equipment

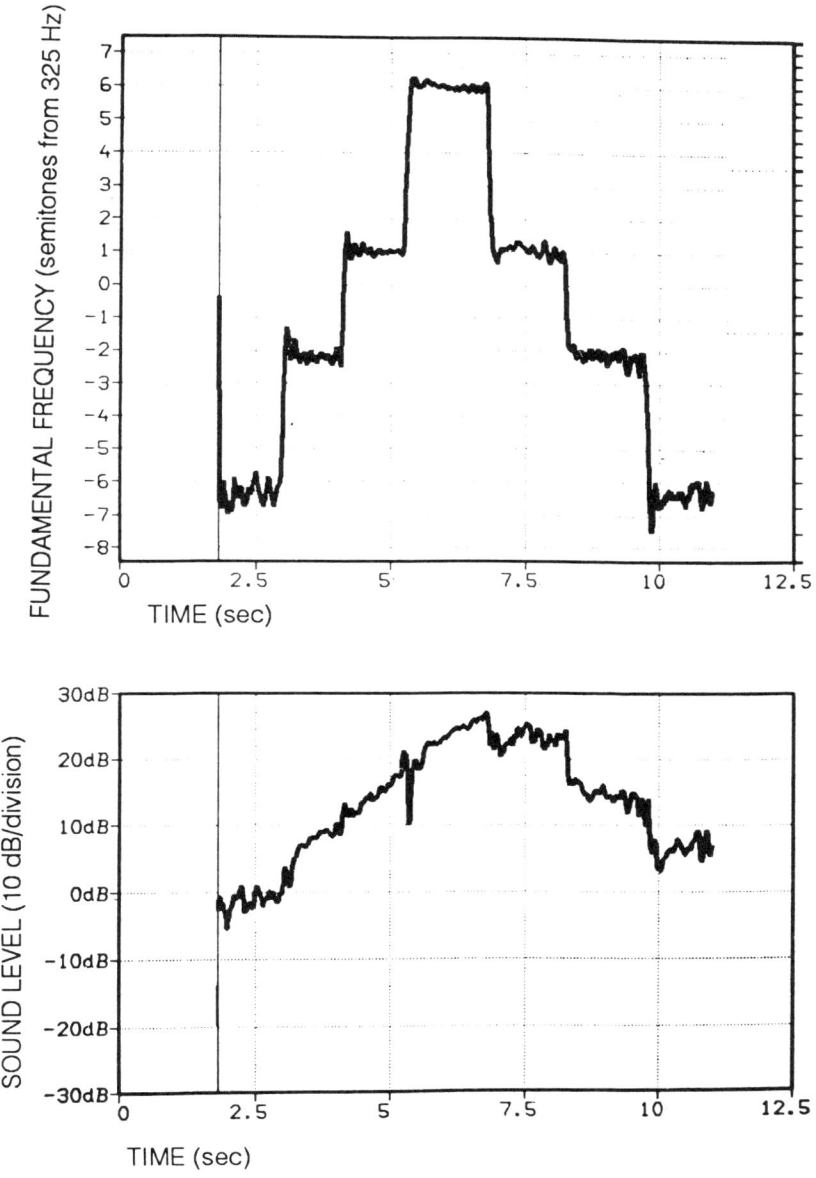

Figure 9.8 Recording by a melody writer program of a female singer performing an ascending and descending triad. Upper panel shows the fundamental frequency; the lower shows the sound level.

208   9. Music and Electronics

The spectrum analyzer can be used for stationary signals, i.e., signals that stay the same. It depicts the spectrum and acts as a level recorder, displaying the sound level in a narrow frequency band that glides in frequency, as the pen moves up the frequency scale. The width of this frequency band is called the *analysis bandwidth*. It determines how much detail one can see in the resulting spectrogram; the narrower the bandwidth, the more details. Analysis with a narrow bandwidth is often rather time-consuming. Figure 9.9 shows an example of the same signal analyzed with two different analysis bandwidths. Spectrum analyzers belong to the standard equipment in acoustic laboratories.

Time-varying complex signals can be analyzed by a special kind of spectrum analyzer. One of them is called *sonagraph*; another is called *voiceprint*. These devices offer a three-dimensional graph. They belongs to the standard equipment of speech laboratories. Time and frequency are represented on the horizontal and vertical axes, respectively, and the sound level is recorded in terms of blackness or colors; the stronger the sound, the blacker the marking. An example of such a recording is shown in Figure 9.10. The sonagram typically shows about 2-sec long parts of the signal, but there are also modern versions that record the sound continuously on a paper roll. The dynamic range shown is generally limited to about 40 dB.

By means of the devices mentioned so far sound signals can be analyzed from a number of different viewpoints. However, the properties of the resonator are often relevant to study. The task then is to find out how the resonator treats sine tones of different frequencies so that the frequency curve of the resonator can be

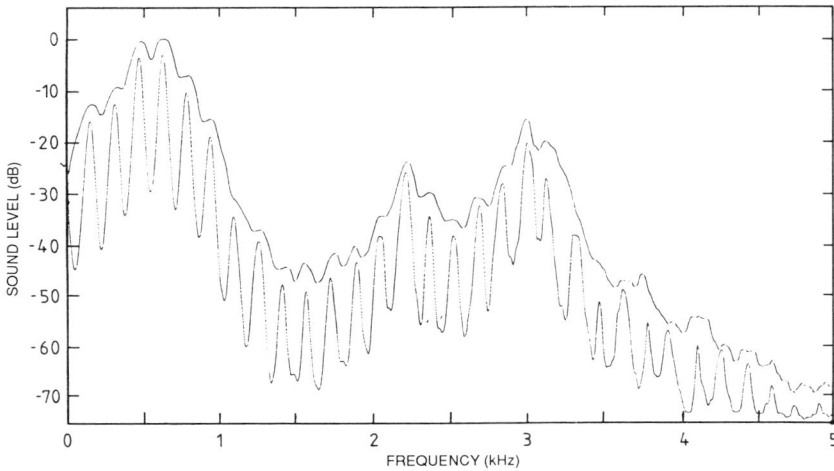

Figure 9.9 Recording of a vowel sound using analysis bandwidths of 10 Hz and 200 Hz.

IV. Acoustic Measurement Equipment

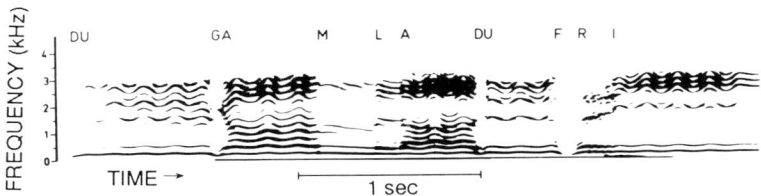

Figure 9.10 Sonograph recording of a song performed by a professional baritone singer.

obtained. The standard method is to mount a sound source providing a sine tone of gliding frequency in one end of the resonator and a microphone in the other. If one wants to examine the input impedance or input admittance, the source and the microphone should be in the same location.

It is rarely possible to adapt loudspeakers in the small resonators of many music instruments, and it is never appropriate, because the vibrations of the loudspeaker slightly change the resonator characteristics. Rather one uses loudspeakers attached to a thin tube, the end of which is inserted into the resonator. The thin tube, however, has resonances of its own, and it is important to keep track of these.

A rather special type of alternative is the so-called ionophone. It works with an electrical discharge between two thin metal wires, some millimeters apart. The discharge generates sound corresponding to the input electrical signal. The advantages are very small dimensions and a practically perfectly straight frequency-response curve. Further it does not disturb the acoustical properties of the resonator.

In analyzing and understanding how acoustic resonators behave, electroacoustic analogies have been found very productive. The point is that the acoustic properties of inertia, compliance, and friction have equivalents in the electrical properties inductance, capacitance, and resistance, respectively. Because of these parallels, resonator properties can be calculated using the same equations as for electrical systems. The reader eager to dive more deeply into these matters is referred to special literature.

Almost all the above devices can be and are also largely replaced by the computer. There are smaller or greater analysis programs available on the market that convert your home computer into a collection of powerful acoustic analysis tools. There are programs displaying fundamental frequency, sound level, or spectrum versus time. Taking their universality into account, computers must be said to be rather cheap, and the prices tend to decrease with time. Computers are completely dominating in all acoustic and phonetic laboratories.

The acoustic signal is then converted to digital form in an ADC, which must meet specially high demands when dealing with music signals. The reason is that the relevant frequency and amplitude ranges are much wider than for speech.

Still, the computer has not totally killed the market for dedicated devices such as level recorders, oscilloscopes, etc. The reason would be similar to the one that makes possible the sale of sawing machines. Although everyone can make a sawing machine from his or her electrical drill, it is sometimes very handy to have access to machines that are always ready to do whatever they can so that one does not have to first make a lot of adjustments to convert it to the kind of tool one needs.

V. SOME WORDS ABOUT RESEARCH METHODS

What, then, is the point of using all these wonderful tools for acoustic analyses of music instruments? Since the early days of music acoustic research after World War II, much endeavor has been spent on exploring the acoustic characteristics of music instruments. Researchers have peeped at music sounds through the more or less narrow keyholes provided by instrumentation for acoustic analysis, and these keyholes have obviously been considerably widened in recent years. The spectral properties of tones from different instruments have been examined, the agreement between the fundamental frequencies used in performance and those represented by equally tempered tuning have been analyzed, and tone onsets have been investigated. Much of what was presented in Chapters 5, 6, and 7 was gathered from such investigations.

The results of such research are valuable as they replace, or at least complement, the traditional verbal descriptions and unconscious knowledge about how music instruments sound; such descriptions suffer from the fatal problem that it is almost impossible for somebody else to know what they mean. For example, what does it mean that a sound is a "somewhat sharp D flat, rather weak, with a muffled timbre and imprecise onset"? However, if we can specify fundamental frequency, amplitude, spectrum, onset characteristics, etc., for a tone, other persons have a chance to find out what that description stands for. Thus, we might say that the acoustic description of a sound offers a possibility of bridging the sound perception gap between different people.

However, an acoustic description is not the end product. It merely answers the question HOW?; it offers an account of raw data, devoid of interpretation. Also, it hides much interesting information as to how the composer, the instrument builder, and the player took the characteristics of the human perceptive system into account, as was explained in Chapter 1. A worthwhile follow-up of the acoustic description of music instrument sounds is to find the reasons WHY.

The explanation can face two different directions. One is the properties of the acoustic systems that generated the sound. The other is the properties of the system meant to receive and process the sound, i.e., human auditory perception.

Let us take a concrete example. As we saw in Chapter 6, the tuning of the piano deviates slightly from the equally tempered tuning; the octaves are tuned some-

V. Some Words about Research Methods

what wider than the 2:1 frequency ratio for a mathematically pure octave. It could later be shown that these deviations could be accounted for by the inharmonicity of the partials present in the struck piano string. If one tunes the piano octaves beat free, they will be slightly stretched. But why does the ear accept stretched octaves? Later investigations of what our hearing perceives as a pure melodic octave showed that such octaves are stretched, both for sine tones and complex tones!

The revealing of these two facts—the inharmonicity of the string and the ear's preference for stretched octaves—has made us wiser. We understand that it is not an error but rather a merit of the piano to depart from mathematically pure octaves. This insight should be of value when new music instruments are being constructed. At the same time, the field of auditory perception has gained new knowledge of interest; when theoretical models are built of the hearing system, this property must fit into them. Music science has also gained another piece in the puzzle revealing the conditions for music experience. Most of this information would, of course, be missing, had researchers stayed content with seeing that—for some mysterious reasons—the piano octaves are slightly stretched.

We stated in Chapter 1 that the first step in the research process—in music science as in any other type of science—is to find the answer to the question HOW? The answer to that question gains considerably in interest when the follow-up question WHY? has been answered. After having read this book this far, the reader may feel a bit disappointed with regard to answers to the latter question. The reason would be that music acoustic research is still a young branch of science; the challenging truth is that only a microscopic fraction of the potential knowledge has yet been revealed. In older fields of science (e.g., medicine), research has reached much farther, so that a great number of phenomena can now be explained. However, thanks to the many new powerful research tools, development is now running swiftly in the music science area, and many questions left open in previous editions of this book (1978) could be eliminated in this edition. And many more will certainly be eliminated in years to come: Music acoustics is a fascinating field!

CHAPTER 10

Music as Communication

I. INTRODUCTION

Music acoustics attempts to describe and explain music sounds. The construction and function of the instruments decide why the sounds possess their characterizing properties, as we have seen before. Another factor of obvious relevance is the properties of auditory perception, as has also been demonstrated.

Yet another factor of obvious relevance is how the music is composed (i.e., its structure). When playing, the player demonstrates an understanding of this structure; for example, he or she signals in certain ways the end of a phrase, and by means of other performance details, the player shows a number of other things: where the harmonies shift, which tones are remarkable and which are self-evident, which belong together, and so on.

It is a huge and fascinating task to study how musicians show their understanding of the music. The first attempts were made in the 1930s by Carl Seashore, but the field was then abandoned almost entirely until the 1960s, when the Swedish musicologist Ingmar Bengtsson started to gather measurements on music performance. During the past decade, questions on musical performance have attracted a fair number of researchers.

One can discern two major forces behind this development of researchers' interest—one quite funny, the other serious. For a number of years typists' skilled behavior attracted the scientific interest of many psychologists. This interest survived until the advent of microcomputers with their word processors, which deprived typing of most of its earlier glamour. Then, the interest swung toward another type of skilled behavior (*viz.* piano playing), certainly a more fascinating topic for the vast majority of people who are musically interested in one way or the other.

The other factor is that in our time, contemporary music is often performed by computers with no other musician involved than the composer. Then, it is valuable to know what the indispensable contributions are that the musicians add to the sound sequences in traditional music.

I. Introduction

The musical importance of these contributions becomes overly evident when music is played exactly as written in the score. For a musician, it is an impossible task to play in exact accordance with the notation in the score. Many of the musicians' additions appear as so self-evident that they are made unconsciously. But beginners and players rightfully regarded as unmusical sometimes offer very striking examples of more machine-like performances.

Still more convincing demonstrations of the musical relevance of the musicians' contributions can be listened to when computers produce a 100% "verbatim" sound translation of the note signs. This opportunity is available in many places these days. Many music programs and synthesizers accept a music score as the input and produce the corresponding sound as the output. The result shows, almost without any exception, that music suffers seriously under such conditions. The music sounds agonizingly meaningless, nagging, mechanical, pedantic, ...

How can one gain an insight into what is between the note signs and the sound? One method has been to study in detail how professional musicians play. For example, many researchers have measured the actual tone durations in performances. By this method, results have been obtained suggesting that musicians apply certain principles according to which they, with or without knowing, add and subtract duration, sound level, and vibrato and make adjustments of fine tuning and timbre when playing.

Another method in music performance research has been to change performance details systematically and listen to the resulting effects. Also in this way, many interesting discoveries have been made in terms of principles that musicians appear to apply when performing music.

What means do musicians have at their disposal for expressing their musical interpretation of a piece of music? One important means is the *exact duration* of the tones. Tone duration may be varied within very wide limits. For example, an eighth note may be as short as a sixteenth note under some conditions and as long as a quarter note under other circumstances. A great advantage with duration is that in a rhythmical sequence, exceedingly small perturbations of the regular pattern produce significant effects: Under some circumstances, 10 msec is enough to evoke an effect. Thus, the difference limen is small, and thanks to categorical perception, duration categories are very wide, actually so wide that they overlap considerably!

However, apart from tone duration, there are many other means available to musicians for conveying their musical interpretation of the piece. Between tones, very short micropauses can be inserted in a meaningful way. Further, the vibrato can be varied in many instruments, as well as the fine tuning of fundamental frequency.

In this chapter we will present some knowledge gained rather recently about this. The reader will notice that the knowledge is not very broad, but also that it is very exciting. The results elucidate how music can work as a communication between player and listener. Not many of these things have yet appeared in textbooks, so the source of information is largely scattered in articles in different journals and proceedings of scientific meetings, and some texts are hard to find. Therefore, a reference list has been added at the end of this chapter.

II. TONE GLUE

One of many fascinating aspects of music is that tones following each other in time stick together, as it were, so that we hear motives and melodies. In other places in the sound sequences, the melodic patterns break. This phenomenon is often referred to as *grouping*: Often very long sequences of sound events appear in our minds as *groups of sounds* or *melodical gestalts*. Some of these groups are referred to as *motives*. The phenomenon that tones appear to belong to different tone groups is also sometimes referred to as *streaming*.

The mechanisms underlying these perceptual effects are not well understood yet, but the phenomenon is robust and highly relevant to music listening. Also, some interesting formal observations have been reported. They concern both the composers' and the players' contributions.

In a dissertation by the Dutch scientist Leon van Noorden (1975), this phenomenon was thoroughly examined, and the terms *fission* and *fusion* were proposed for the cases that adjacent tones sound as if they belong together or form parts of different gestalts respectively.

Some striking results are shown in Figure 10.1. It shows under what circumstances tones merge to melodic patterns and when the parts fall apart. The figure illustrates the fact that tones can be glued by two different kinds of paste. One is *pitch proximity:* Tones that are close in pitch tend to stick together to build melodic patterns. Another kind of tone paste is *duration*. Even if the interval between two adjacent tones is wide, these tones can still merge to build a melodic pattern if they are long.

In the figure we can see that van Noorden worked with very quick tone passages, but the two types of glue mentioned also work in tempos more typical of melody playing. The figure shows that there is a neutral zone between the fission and fusion zones. There, both effects can occur. In the furious tempos represented in the figure, fusion occurs only if the pitch interval does not exceed two or three semitones. In somewhat less-rushed tempos, melodic constructions survive somewhat wider intervals.

This phenomenon is well known and has been in frequent use in music for centuries. The term used in music theory is actually *melodic fission*. It depends not only on tempo and interval sizes, but also on timbre.

A German priest and organ theoretician, Ernst Karl Rösler, made an interesting contribution in this area (Rösler, 1952), although his work left undeservedly

II. Tone Glue

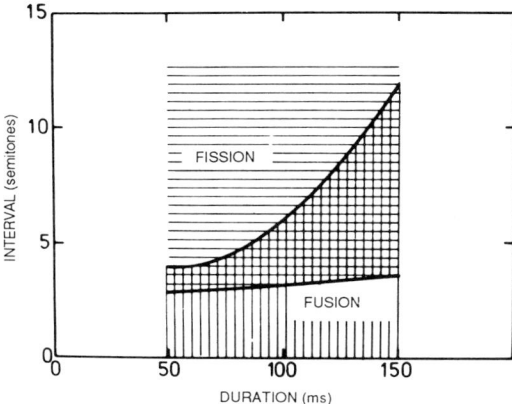

Figure 10.1 Combinations of melodic intervals and tone durations that produce fusion and fission of melodic tone groups or motives, i.e., decides whether the tones are perceived as belonging to the same melodical gestalt. (From van Noorden, 1975.)

shallow traces in the theory of timbre. He introduced the terms *spatial line strength* and *harmony strength* (German *"Raumlininestärke"* and *"Harmoniestärke"*). They refer to certain timbral characteristics that can typically be observed in some organs. In Baroque organs the stops tend to have high spatial line strength; therefore, they are apt to keep melodies together even when the music contains rather wide intervals. Further, such stops treat contrapuntal music in a friendly way, because they make the different simultaneously playing voices easy to track for the ear. In romantic music from the late 19th century, it is often much more rewarding to pay attention to the harmonies than to look for the often nonexisting counterpoint; hence, the organ stops of the romantic organs were given great harmony strength, furthering fusion of simultaneous tones to chords.

Harmony strength and spatial line strength are opposites. Therefore, harmony strength produces melodic fission when wide intervals appear in a melody. And if one plays a simple four-voice choral on an organ with great spatial line strength, the risk is that, for example, the alto voice jumps out of the harmonic structure, presenting itself as a solo voice, just to reveal, without shame and mercy, its lack of contrapuntal excitement.

The effects that Rösler named and discussed, of course, go back to spectrum properties. One might suspect that they relate to the frequency range of the spectrum, perhaps the width of the excitation area on the basilar membrane; if two tones have a similar excitation area, they may possess spatial line strength.

The American psychologist David Wessel further developed the idea that spectral similarity is connected to grouping. He demonstrated that a melodic line crumbles if its tones are played one by one on instruments with very different timbres; the timbral differences were accounted for in a compact way in terms of few timbral dimensions. An account for this interesting phenomenon and a sound example can be found in Pierce (1983). If, however, the melody is played so that adjacent tones are reasonably similar in timbre, there is no great difference in the timbral dimensions between adjacent notes, the melody holds together.

When synthesizing singing, the author has learned that the singer's formant, discussed in Chapter 5, is another important means of keeping tones together. If it does not occur in one tone in a melodic line, that tone appears to fall out of the melodical frame so that the legato is ruined. The singer's formant seems to serve the purpose of a uniform cap, in which the singer needs to dress all tones. The effect is that we understand that these tones belong to the same crowd. The singer's formant appears to be another kind of tonal glue.

Timbral similarity is not the only means to produce a tone grouping. Also its opposite, timbral contrast can serve the same purpose. A typical example is shown in Figure 10.2 showing a long-term average spectrum of two organ stops, one intended for playing the solo voice and the other for the accompaniment in a small organ. Such spectra show the average sound level in various frequency bands. The figure shows in a striking way how the two spectra complement each other: The

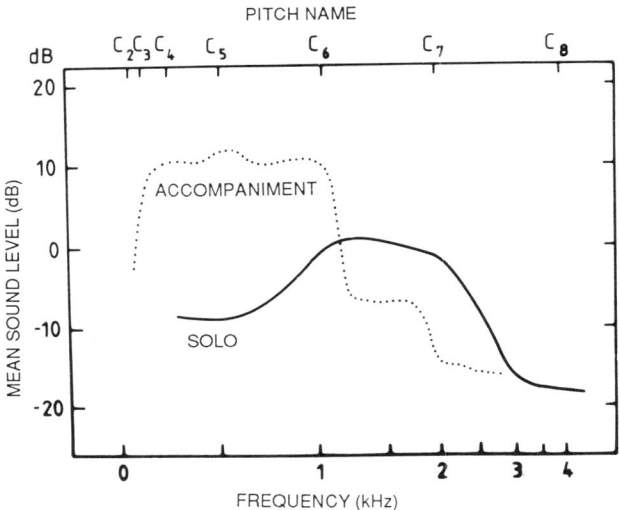

Figure 10.2 Long-term average spectra of two organ stops (Vox Virginea 8′ and Gedakt 8′) intended for playing solo and accompaniment respectively in a small organ. Partials of the accompaniment stop are strongest in the low-frequency range and weak in the high-frequency range, whereas the opposite applies to the solo voice. (From Sundberg and Jansson, 1975.)

II. Tone Glue

accompaniment stop is strong in the low-frequency region, whereas the solo voice is strong in the high-frequency region.

In instruments allowing a continuous variation of loudness, i.e., most instruments except the organ, there is another opportunity to glue tone sequences together. By synthesized sound examples, it has been demonstrated that the tones in a melody can be pasted together by making all tones members of the same long-term sound level change (Sundberg, 1978). Adjacent tones were glued together to a phrase because they formed parts of a common crescendo-decrescendo. Similarly, if adjacent tones differ in sound level, melodic fission is likely to result; however, this effect can be neutralized if steps in the sound level curve are smoothed by making crescendos or diminuendos during the individual tones. In a corresponding way, pauses are used to demarcate boundaries between different parts of a melody, i.e., to create the opposite of affinity.

Musicians insert very short pauses, or micropauses, between certain tones and seem to adjust their lengths very carefully. Often they talk about this as *articulation,* and we will return to this in a moment.

The decay of the tones is naturally significant to articulation. Most instruments contain resonators, and in these the decay is not instantaneous, as we saw in Chapter 2. Particularly in instruments with poorly damped resonators, such as harp and tympani, the tones die away rather slowly. In such extreme instruments, additional damping is provided by putting the hand on the resonator when needed. But also in instruments with a short decay, the exact length of the decay is important, particularly as regards the legato. If one wants to attain a legato in electroacoustic systems, it is necessary to have the tones overlap slightly in time.

Some tones almost appear to become a bit sticky or repellant, depending on the musical context; some tones seem to fit very well when played in succession, whereas others appear to fit poorly and tend to repel each other and break the melodical continuity. In a series of experiments, the American psychologist Carol Krumhansl and co-workers studied this phenomenon (see, e.g., Krumhansl, 1987, 1990). Her experiment was to have listeners first hear a scale and then a probe tone. The task was to rate how well the probe tone served as a continuation of the scale. She also repeated the same experiment with a cadence replacing the scale. Figure 10.3 shows the results. It is evident that tones belonging to the scale received high ratings and vice versa. Also, the members of the tonic chord are rated higher than the nonchord scale tones, whereas nonscale tones are rated low. In yet another group of experiments, she replaced the scale by a cadence and the probe tone by a probe chord and received a grading of the chords. When she later made similar experiments with 12-tone series instead of scales, the results came out quite differently (Krumhansl *et al.*, 1987). These experiments demonstrate that there is a strong component of *expectations regarding the continuation* in music listening and also that these expectations differ when we listen to different musical styles.

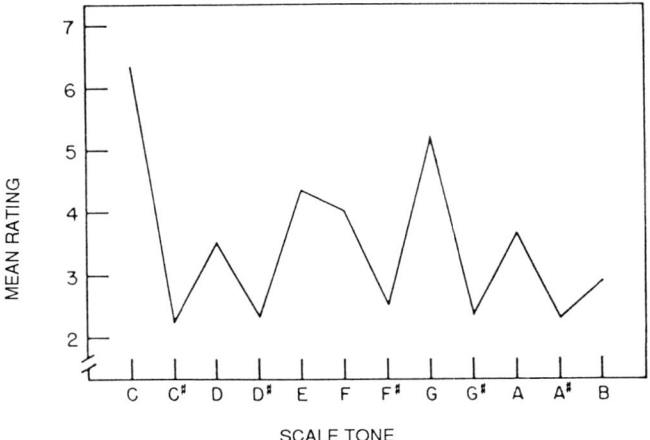

Figure 10.3 Mean ratings reflecting how well an average listener thinks a given tone fits as a continuation of a major scale: The tone G fits very well and received a high rating whereas the tone C sharp and F sharp received very low ratings. (From Krumhansl and Kessler, 1982.)

We have now considered a kind of melodical stickiness, which causes successive tones to cling to each other, so that they form melodic groups, motives, or short melodies. There is another type of tonal stickiness, which causes simultaneous tones to stick to each other so that we hear them as emanating from the same source, i.e., are members of the same spectrum. Indeed, it is quite remarkable that a trained musical ear can hear exactly what tones are played by the different members of, for example, a wind instrument ensemble; what is offered to the ear is actually nothing but a chaotic bunch of partials. How can we hear that these partials came from the oboe and those from the flute, etc.?

Steve McAdams (1982) studied this ability experimentally. He found that a common modulation of the fundamental frequency, however small it may be, is utterly revealing: Those partials, which move according to the same frequency pattern, clinch together and are heard as originating from the same source. Vibrato therefore has a strong effect in this regard. A common change of sound level is another indication. Sometimes a common origin may be revealed by an asynchrony of attacks: If one instrument arrives at a new chord somewhat earlier than the fellow tones, we tend to hear that its partials stem from one source. This important perceptual principle has been called *the principle of common fate*: Those parts that obey a common fate are assumed to belong together. Another cue is the *direction* of the sound in space, in case it is possible to locate the sounds.

As soon as we have discerned a particular voice in a harmonic structure, it seems easier to hear how it continues. For this ability it would be important that we really do not need to hear, in a physical sense, a tone continuously to perceive it as continuous. It is enough to hear a glimpse of it every now and then, about three times per second. Expectations seem highly relevant; it is always easier to discern a well-known than an unknown melody in a polyphonic structure, e.g., presented as a cantus firmus. Many of us have probably also noted that it is much easier to track the viola part in an orchestral structure if we can look at the score while listening.

III. A SOUND MEANS SO MUCH!

A question that has puzzled many people is to what extent a sound may mean particular extramusical things, so that music communication may be used somewhat as a language. During some periods of the history of music, people have been fully convinced that music can mean concrete extramusical things, and one has composed accordingly. This type of music has been referred to as *program music*. This is not to say that the musical value of the piece is dependent on whether the listener understands what the music means. On the contrary, the compositional technology used for a composition is often of very little relevance to the listening experience. At least good program music can be enjoyed without paying any attention whatsoever to what it was supposed to mean.

Actually, this has happened several times in music history. For example, how much does it mean to our enjoyment in listening to the music of Sebastian Bach that he was eager to compose the number 14 into every possible angle of his wonderful musical architecture?

This is not to say that sounds and sound sequences may never represent some very specific, extramusical object. In such cases, a particular sound event is used as a sign of a particular object or relation, i.e., sound is used as a code. Next we will see some examples, and in all these examples, it is evident that we have learned the code in one way or the other.

One type of significance of a sound is its origin. If we hear the sound of tympani, many of us would tend to think of tympani, and if we hear something sounding like footsteps on gravel, we would be prone to think of footsteps on gravel. Here auditory experience taught us the code.

Sound and sound sequences may also act as small trucks loaded with associations that drive into the listeners' heads and unload their cargo. The sound of an organ would bring many minds in a churchy direction, and seagull screams would tend to bring lakes and ocean shore into our thoughts.

Some of these cargos of associations get to stick to the sounds for very good reasons. Other sounds get loaded for less good reasons, if not by pure accident, and such associations are, of course, less interesting. Associations that are hard to wash out from the sound seem more inspiring.

An example of a seemingly very stable association is that of minor tonality and a sad ambience. It seems possible to get rid of it, but lots of strong antidote would be required.

Similarly, certain tempi seem very strongly associated with certain moods. The author had the exciting experience of meeting a very unusual and skilled physician, who cured blood circulation problems by a type of vacuum pump. This pump alternately raised and lowered the pressure around the part of the body that suffered from poor circulation, and in this way circulation improved. He told me that a slow pumping rate caused the patients to become sleepy, so he had to serve them lots of coffee to keep them awake. If the pump rate was too quick, on the other hand, his patients became agitated and needed to be calmed down.

These effects of tempos are perhaps not completely impossible to understand. We all have certain innate frames of reference for breathing, pulse rate, and footsteps, and these rates are often quite revealing in terms of the underlying mood. Life taught us the code.

There are not only tempos that may be eloquent. It often happens that tone sequences resemble gestures. Gestures can be very eloquent and in this sense they may "mean" quite special things. Further, our voice organ provides us with an ever-present translation of gesture to sound patterns: Every change of a voice sound reflects a certain change in one or more of the different parts of the voice organ. For example, a pitch change generally reflects a change in the vocal fold length accomplished by a contraction of certain muscles, which regulate the positioning of the cartilages holding the vocal folds. We may say that we practice our ability to interpret sound sequences in terms of moods every day as soon as we talk or listen to speech. Again, then, life taught us the code.

Sound and sound sequences may also mean something particular because they allude to or are similar to something familiar. If a melody may sound lamenting, it is perhaps in some salient way similar to, i.e., partially identical to, a lamenting speech. The Australian psychologist Manfred Clynes has studied the relations between gesture, sound, and emotional expression in an interesting and unusual way (Clynes, 1983).

Some meanings seem to be very loosely fastened to their sounds. Program music often generously provides examples of meanings that are almost impossible to detect. For instance, an ascending scale followed by a descending scale may simply mean "church tower," but few people would have the pleasure of enjoying this while listening to the music.

During the Baroque era, composers worked with what in German was called *Affektenlehre*, which could perhaps be translated as "affectology." According to this idea, certain intervals were associated with certain things. Probably the expert listeners such as composers of that time could experience these meanings of details in the compositions while listening. For other people this would be hard, though, because the code is not taught by life but rather by special education.

Sound and sound sequences may also acquire meanings by means of definitions and/or traditions: The composer teaches the listeners the code. If a particular theme always is presented when a certain person enters the stage in an opera, this theme will soon come to "mean" this person. The composer teaches the listener the code. Richard Wagner certainly used this trick systematically in his operas. In this sense the Tannhäuser theme means Tannhäuser for many Wagner-loving listeners.

Some types of meaning can be quite stable. For instance, *cadence* and *hemiola* are two compositional means that serve particular functions such that they mean particular things. After centuries of serving a particular purpose, namely, to mark the end of a structural unit, they have received a structure-marking effect, and certainly most music listeners are acquainted with this kind of musical meanings. However, this is a slightly different form of meaning, because this meaning is intramusical, whereas the ones previously discussed were extramusical, the music referring to something outside itself. Still the code has to be learned. In this case, listening to music taught us the code.

IV. PERFORMANCE RULES

We just mentioned that a verbatim realization of the music score, faithful down to the last millisecond, does not result in a musically acceptable performance. Rather, the performance sounds machine-like (which it, of course, is!). This shows that musicians perform music such that, for example, the durations of the tones deviate from their nominal values in the score. These deviations are often called *expressive*. They are sometimes minute, sometimes great, but it seems that an attentive listener always can hear them in some sense. But first of all they are meaningful. For this reason we will call them *meaningful deviations*.

Thus, these meaningful deviations cannot be replaced by random variations: This is not very astonishing as it really means nothing more than that musicians cannot be replaced by random functions. These meaningful deviations from the nominal description given by the score represent an important and interesting part of the science of musical sounds. And, of course, they are far more interesting than the meaningless deviations, which occur by accident, when the musician fails to execute his or her intentions when playing.

Communication obviously requires that both sender and receiver, i.e., player and listener, are acquainted with the same code. Three questions present themselves with respect to this code:

(1) What is this code?
(2) What information does it convey?
(3) Who taught the listener the code?

These questions are basic in music acoustics, and their answers deepen our understanding for what music and musicality really imply. In this section we will look a bit more closely at what is presently known about the principles that underlie the performance of music and listening to music performance. The results are rather fresh, and for this reason they are far from being final and exhaustive.

It seems clear that the meaningful deviations are made according to certain principles, and at present it is difficult to see how general these principles are. However, it is obvious that a musician is by no means free to invent any principles he or she would like, because listeners would be unable to interpret new and freely invented deviations and would hence find them meaningless, i.e., a sign of poor playing technique. Next we will present some of these principles or rules and speculate as to their musical function. Some rules have emerged from various researchers' measurements on music performance. Some are fruits of the author's cooperation with a professional violinist, Lars Frydén.

The method in the latter research has been to have the professional musician "teach" a computer how to improve its playing of music examples. The computer is programmed to convert input note signs to the corresponding sound sequences, using a MIDI synthesizer as the output device. In this program, context-dependent rules are introduced that lengthen and shorten notes, create crescendos and decrescendos, accents, micropauses, etc. By listening carefully to how the performance is affected by different rules, the musician has been able to tell how the rule should be changed or complemented to produce a better performance. This method is quite common in music acoustics as well as in many other areas of scientific research. It is called *analysis by synthesis*.

The rules appear to serve two main musical purposes: differentiation of pitch and duration categories, and grouping of tones that musically belong together.

The differentiation is facilitated by enhancing the differences between categories. For example, high notes are played somewhat sharp and low notes are played somewhat flat, thus increasing the differences between the pitch categories. At the same time, the octaves are a little stretched. Further, short notes are played a little shorter than normal and long notes a bit longer. Another example is that the difference between dotted quarter note plus eighth note in two-fourth time, and quarter note plus eighth note in three-eighth time, is increased by taking some duration from the long note and adding it to the following shorter in three-fourths (or halves) time. Thus, in this particular durational context, the short notes are lengthened.

Most music theorists agree that music structure is hierarchical. This means that small units form greater units, which together with other units form still greater units, which together with other units form still greater units, which together with other units form still greater units, and so on. For example, a couple of verse-foot motives form bars, which together with other bars form pairs of bars, which together with other pairs of bars form subphrases, which together with other subphrases form phrases. Most music seems to be composed in this way.

We may insert at this point that music is by no means exceptional in this regard; rather humankind seems to leave a hierarchical structure as something like a fingerprint on all creations: architecture, constitution, jurisdiction, administration,

IV. Performance Rules

literature, speech, etc. Perhaps the hierarchical structure reflects an essential property of the human mind.

When listening to computers playing music, one almost tends to be offended by its total lack of musical understanding, and part of what is felt to be missing is the phrasing. When music is performed without consideration of the phrase structure, the result sounds overly stupid and insensitive. This suggests that we are used to hearing phrase marking in performance, apparently because musicians always insert them in their performances. This marking of structure occurs at different structural levels.

Phrase endings are marked by a lengthening of the final note or notes. In measurements of music performance, a trend to accelerate the beginning of new phrases has also been observed. Subphrase endings are often marked by inserting a micropause after the final note. An alternative way of marking structural boundaries is to insert a contrast in timbre or sound level.

If we continue our descent in the hierarchy to the bar level, typical bar patterns have been found. In the Vienna Waltz, Bengtsson and Gabrielsson (1983) found that the first and second beats took, on the average, 25% and 42% of the bar duration rather than the nominal 33%. In other words, the first beat is short and the second is long. In a similar way, Clynes advocated the assumption that different composers must be played with composer-specific durational bar patterns. This is probably true although not for all music. For example, many composers, particularly during the Baroque and Classic eras, often play little musical games by moving the perceived bar lines around so that they do not always match those given in the notation. Also, grouping occurs not only within but also across bars. Some rules move tones belonging together closer in time and vice versa, and this would apparently violate a rigid durational bar pattern.

The hierarchical structure is not the only thing that music has in common with other types of manifestations of the human mind. Also the phenomenon of predictability plays an important role in music as well as in other types of well-functioning communication systems. In speech, for example, predictability seems to play an extremely important role, and it obviously varies during the message, depending on what is being said. Probably, it is owing to predictability that speech communication rather successfully survives all these beats and shocks in terms of momentary masking noises that in real life typically interweave speech communication. When we listen to speech we are rather skilled at filling the gaps in the signal flow caused by disturbances. Likewise an attentive music listener would be capable of pretty well predicting what chords and pitches will occur next. For instance, it is not all that impossible to tell if a person played the wrong note by mistake, even when we have never heard the piece before. Some of the expectancies underlying this ability may be driven by universal principles, such as reasonable continuity. However, predictions are certainly based on familiarity with the musical style.

Experiences from the computer's way of performing music in a one-to-one relationship with what is nominally written in the score clearly demonstrate the importance of taking into account what is expected and what is unexpected in the music. Examine Figure 10.4. It shows a theme from Schubert's B minor ("Unfinished") symphony. The theme is in D major. All of a sudden, a D sharp appears, precursor of a visit in an E minor harmony. This D sharp is by no means equal to the other notes; rather it's appearance is nothing less than sensational. If performers fail to show that they in fact realize what is trivial and what is sensational, their performance sounds terribly unmusical.

If one examines the underlying harmony in the example shown in the figure, further support is found for the interpretation above. The D sharp is part of a B major chord, and this chord is no less than sensational, too. It does not suffice to act as if nothing were the matter, when this magnificent chord appears in the music.

Thus, it appears that there is a predictability that varies with time in music, and it is necessary that this is reflected in the performance. This, again, is very similar to what happens in speech. It is easy to get somewhat upset if while listening to a text read aloud, the reader arrives at something utterly important but neglects to emphasize it, e.g., by increased syllable length, voice pitch gestures, or so. It appears as if the human perceptual system expects this. A secret communication seems to exist not only between the composer/author and the listener, but also between the performer/narrator and listener. The latter communication seems to contain the message "You and I both realize that this is important while that was to be expected!"

Needless to say these effects should not be exaggerated. The good musical judgment of the talented musician leads the way.

It is also interesting to see in what way and when emphasis and its opposite deemphasis are used in music. What is it that needs emphasis? To predict this with respect to pitch classes, the old classical circle of fifths turns out to be a key. The farther away from the root of the chord a tone is located on the circle of fifths, the more emphasis it appears to need in the performance. A close cousin of this distance has been called the *melodic charge* of the tone, which is very similar to the ratings that the different tones received in Krumhansl's (1987) experiments

Figure 10.4 Second theme with underlying chords from the first movement of Franz Schubert's B minor symphony, D 759 (the "Unfinished"). Chords are given in terms of the interval, in semitones, between the root of the tonic and the root of the chord; thus, 0 means tonic (I) and 7 means dominant (V).

IV. Performance Rules 225

described above. Moreover, the higher the melodic charge of the chord notes, the higher the harmonic charge of that chord. The higher the melodic charge of a tone, and the higher the *harmonic charge* of a chord, the less expected they appear to be and the more emphasis they need in performance.

A melodically charged note is lengthened and is played louder and, if possible, with more vibrato. Another way to give emphasis is, at least in piano playing, to insert a micropause, delaying the appearance of the remarkable event somewhat. When an increase of the harmonic charge is approaching in a chord sequence, a crescendo appears to be appropriate and a decrescendo seems needed in the opposite situation. This gives us an idea as to how the musician can use emphasis to stress and perhaps in some sense also to comment on melodic and harmonic events while playing.

Obviously, the above applies to tonal music. In atonal music it is inappropriate to talk about chords and roots of chords. Here it seems more adequate to introduce the notion of chromatic charge, defined such that it increases when the successive tones approach each other in pitch, and vice versa. When the chromatic charge increases in a piece, the same performance events seem appropriate as when the harmonic charge is increasing in a harmonic progression in tonal music.

Above we have seen how and when the musicians assist the listeners' digestion of the musical signal flow by applying differentiation, emphasis, and grouping rules during playing. As the described research results accurately define how this is realized in the performance, it is also possible to examine the code used. Thus, by what means are emphasis and grouping demonstrated in music performance?

Phrase endings are marked by a lengthening of the final tone or tones, and subphrase endings are marked by micropauses. This is, in fact, exactly the same code as is used in speech! In sentences the last syllables are lengthened, and small pauses are inserted after smaller groups of words. Also with respect to emphasis marking, there are strong parallels with speech. Important events are lengthened in both cases. In music not much can be done with pitch, but the depth of the vibrato modulation can be increased, provided the instrument allows it.

These parallels are certainly most interesting. They suggest that in listening to music, a competence is used that we acquired when we learned to understand speech. To be musical does not seem to be all that exotic, even when seen from an unmusical viewpoint!

Speech and music apparently share a good deal of the code for expression. This is not to say that music can use only codes that appear in speech. There may be many reasons why a code used in speech communication is not appropriate for music communication. Still, the number of possibilities is, of course, far from being infinite whether in speech or in music. The obvious reason is that *no code will be successful if it is not common to the sender and the receiver.*

The code used in speech can hardly be the origin of that used in music, even if they are partly identical. In the case of the final ritard in music, announcing the coming termination of some pieces of music, striking similarities have been found between the way in which the time interval between the tones increase and the way in which the intervals between the footsteps are increased when a person stops running. We must admit that the idea of introducing a final ritard at the end of a piece is very smart.

Above we have mentioned differentiation, marking of emphasis, and grouping. This, of course, does not exempt the list of meaningful deviations occurring in music performance. In music performance it is also mandatory to create the proper emotional ambiance. How this happens is not well understood. It seems reasonable to assume that the code originates from speech and gestures, or from the "body language," as many people would like to call it. As an example it is again appropriate to refer to a well-known experience in attempts to have the computer play musically: Quick changes of sound level always sound aggressive or pushy, whereas such changes sound peaceful and friendly if they are slow. The relations between gesture and expression in speech and music have been eloquently elaborated on by the Hungarian phonetician Ivan Fonagy (1981).

As was mentioned before, we have only sketched a fragmentary picture of what it is that makes music worthwhile to listen to. Research has not come all that far, but hopefully it will make quick progress in the future. The topic is undoubtedly fascinating, and it gives deep insights into the basic requirements for music communication.

V. BIBLIOGRAPHY

Finally we will give a short list of references in this chapter, because, contrary to the areas treated in earlier chapters, there are no exhaustive texts overviewing the entire topic.

Bengtsson, I., and Gabrielsson, A. (1977). Rhythm research in Uppsala. *In* "Music Room Acoustics." Royal Swedish Academy of Music, Publication Nr **17**, 19–56.

Bengtsson, I., and Gabrielsson, A. (1983). Analysis and synthesis of musical rhythm. *In* "Studies of Music Performance" (J. Sundberg, ed.). Royal Swedish Academy of Music, Publication Nr **39**, 27–60.

Clarke, E. Categorical rhythm perception: an ecological perspective. *In* "Action and Perception in Rhythm and Music" (A. Gabrielsson, ed.). Royal Swedish Academy of Music, Publication Nr. **55**, 19–47.

Clynes, M. (1983). Expressive microstructure in music, linked to living qualities. *In* "Studies of Music Performance" (J. Sundberg, ed.). Royal Swedish Academy of Music, Publication Nr **39**, 76–181.

Edlund, B. (1985). "Performance and Perception of Notational Variants." Almqvist & Wiksell, Stockholm.

Deutsch, D. (1982). *In* "The Psychology of Music." Academic Press, San Diego.

Dowling, J., and Harwood, D. (1986). "Music Cognition." Academic Press, New York.

Fonagy, I. (1981). Emotions, voice, and music. *In* "Research Aspects of Singing." Royal Swedish Academy of Music, Publication Nr **33**, 51–79.

V. Bibliography

Handel, S. (1989). "Listening." MIT Press, Cambridge, MA.
Krumhansl, C. (1987). Tonal and harmonic hierarchies. *In* "Harmony and Tonality" (J. Sundberg, ed.). Royal Swedish Academy of Music, Publication Nr **54**, 13–32.
Krumhansl, C. (1990). "Cognitive Foundations of Musical Pitch." Oxford University Press, New York.
Krumhansl, C., Sandell, G., and Sergeant, D. (1987). "The perception of tone hierarchies and mirror forms in twelve tone serial music." *Music Perception*, **5**, 31–78.
McAdams, S. (1982). Spectral fusion and the creation of auditory images. *In* "Music Mind and Brain" (M. Clynes, ed.). Plenum Press, New York.
van Noorden, L. (1975). "Temporal Coherence in the Perception of Tone Sequences." Institute for Perception Research, Eindoven.
Rösler, E. (1952). "Klangfunktion and Registrierung." Bärenreiter, Kassel.
Shaffer, L. (1981). Performance of Chopin, Bach, and Bartok: studies in motor programming. *Cognitive Psychol.* **13**, 326–376.
Sloboda, J. (1983). The communication of musical metre in piano performance. *Q. J. Exp. Psychol.* **A35**, 377–396.
Sundberg, J. (1978). Synthesis of singing. *Swed. J. of Musicol.* **60:1**, 107–112.
Sundberg, J. (1988). Computer synthesis of music performance. *In* "Generative Processes in Music Performance" (J. A. Sloboda, ed.). Oxford Science Publications, Oxford.
Todd, N. (1985). A model for expressive timing in tonal music. *Music Perception* **3**, 33–58.

References

Askenfelt, A. and Jansson, E. (1988). From touch to string vibration—the initial course of the piano tone, *J. Acoust. Soc. Amer.*

Alonso Moral, J. & Jansson, E. (1982). Input admittance, Eigenmodes, and quality of violins, Speech Transmission Laboratory Quarterly Progress and Status Report 2–3/1982, 60–75.

Backus, J. (1969). "The Acoustical Foundations of Music," W.W. Norton and Co., New York

von Békésy, G. (1960). "Experiments in Hearing," McGraw-Hill, New York.

Benade, A. H. (1976). "Fundamentals of Musical Acoustic," Oxford University Press, New York.

Björklund, A. & Fintoft, K. (1966). "Elementaer musikalisk akustikk," Aschehough, Oslo.

Chowning, J. M. (1977). The synthesis of complex audio spectra by means of frequency modulation, *J. Audio Eng. Soc.* **21**, 526–534.

Clarke, E. (1987). Categorical rhythm perception: an ecological perspective. *In* "Action and Perception in Rhythm and Music," (A. Gabrielsson, ed.), K Musikaliska Akademiens Skriftserie Nr 52, 19–33.

Dekan, K. (1972). Auswertung von musikalischen Dynamikbereichen bei verschiedenen Blechblasinstrumentspielern und die Klanngfarbenänderungen bei piano, mezzoforte und forte (h. Glahn, S. Sörensen and P. Ryom, eds.), Report of the 11th Congress of the International Musicological Society in Copenhagen, Vol. I, 351–355.

Denes, P. B. & Pinson, E. N. (1975). "Det talade språkets kommunikationskedja," Wahlström & Widstrand, Stockholm.

Fant, G. (1958). Modern instruments for acoustic studies of speech, *Acta Polytechnica Scandinavia Ph1*, 246/1958.

Fletcher, N. H. & Rossing, T. D. (1991). "The Physics of Musical Sounds," Springer Verlag, New York.

Gade, A. (1986). Acoustics of the orchestra platform from the musician's point of view, *In* "Acoustics for Choir and Orchestra," (S. Ternström, ed.), K. Musikaliska Akademiens Skriftserie Nr 52, 23–42.

Hadding, K. & Petersson, L. (1972). "Experimentell fonetick," Gleerups, Lund.

Hagerman & Sundberg (1980). Fundamental frequency adjustment in barbershop singing, *J. Research in Singing* 4–1, 3–17.

Hall, D. (1980). "Musical Acoustics," Wadsworth, Belmont, California.

Handel, S. "Listening. An introduction to the perception of auditory events," The MIT Press, Cambridge, Massachusetts.

Jansson, E. (1977). "Kompendium in musikakustik, Inst Talöverföring & Musikakustic," KTH, Stockholm.

Jansson, E. V. & Alonso Moral, J. (1980). Recent violin research at KTH, New York: Sound Generation in Winds, Strings, Computers," K. Musikaliska Akademiens Skriftserie Nr 29, 229–245.

Krumhansl, C. (1987). Tonal and harmonic hierarchies, *In* "Harmony and Tonality" (J. Sundberg, ed.), Royal Swedish Academy of Music, Publication Nr 54, 13–32.

Krumhansl, C. & Kessler, E. (1982). Tracing the dynamic changes in perceived tonal organization in a spatial representation of musical keys, *Psychol. Rev.* **89**, 334–368.

Krumhansl, C. (1990). Cognitive Foundations of Musical Pitch, Oxford Psychology Series Nr 17, Oxford University Press, New York.

Martin, D. & Ward, D. (1961). Subjective evaluation of musical scale temperament in pianos. *J. Acoust. Soc. Amer.* **33**, 582–585.

Meyer, J. (1972). "Akustik und musikalische Auffuhrungspraxis, Das Musikinstrument," Frankfurt a M.

van Noorden, L. (1975). Temporal Coherence in the Perception of Tone Sequences, dissertation, Techn. University, Eindhoven.

Olson, H. F. (1968). "Musical Engineering." McGraw-Hill, New York.

Plomp, R. & Levelt, W. (1965). Tonal consonance and critical bandwidth. *J. Acoust. Soc. Amer.* **38**, 548–560.

Podlesak, M. & Lee, A. (1988). Dispersion of waves in piano strings. *J. Acoust. Soc. Amer.* **83**, 3057.

Risset, J. C. & Mathews, M. (1969). Analysis of musical instrument tones, *Physics Today* **22**, 23–30.

Roederer, J. G. (1979). "Introduction to the Physics and Psycho-physics of Music," Springer-Verlag, New York.

Rossing, T. (1982). "The Science of Sound," Addison-Wesley Publishing Co., Reading, Massachusetts.

Schuck, O. H. & Young, R. W. (1943). Observations on the vibrations of piano strings. *J. Acoust. Soc. Amer.* **15**, 1.

Shackford, C. (1961 and 1962). Some aspects of perception. I, II and III, *J. Music Theory* **5**, 162–202; **6**, 66–90; and 2953.

Pens, E. E. (1970). Elementa in akustisk fonetik, Stockholms Universitet: Institutionen för fonetik.

Stensson, K. (u å). Banda bättre, 3M Company, Stockholm.

Stevens, S. S. & Davis, H. (1938). "Hearing," John Wiley & Sons, New York.

Sundberg, J. (1966). "Mensurens betydelse in öppna labialpipor." Stockholm: Almqvist & Wicksell.

Sundberg, J. (2: 1986). Röstlära, Stockholm: Proprius.

Sundberg, J. & Jansson, E. (1975). Long-time-average spectra applied to analysis of music, Part II: An analysis of organ stops, *Acustica* **34**, 269–274. (Fig 10.2)

Taylor, C. A. (1965). "The Physics of Musical Sounds," The English Universities Press, London.

Ternstrom, S. & Sundberg, J. (1988). Intonation precision of choral singing, *J. Acoust. Soc. Amer.* **84**, 59–69.

Weinreich, G. (1977). Coupled piano strings, *J. Acoust. Soc. Amer.* **62**, 1474–90.

Zwicker, E. & Feldtkeller (1967). "Das Ohr als Nachrichtenempfänger," S. Hirzel Verlag, Stuttgart.

Ågren, K. (1976). Alto and tenor voice and harmonic intervals between them (in Swedish), Unpublished theses work, Department of Speech Communication and Music Acoustics, R. Inst. Technology, Stockholm.

Index

A

absorbing surface, 174–176
absorption, 171–174
absorption coefficient, 172
absolute pitch, 51–52
accelerometer, 206
accordion, 128
aeroplane, 41
aftertouch, 203
Affektenlehre, 220
age, effect on hearing, 54
Ågren, K., 102
air absorption, 171, 173
air pressure, 8
air reed instruments, 108, 137–140
alto singer, 118, 126
amplifier, 154, 193
amplitude, 10, 11, 12
analog signal, 198
analog-digital-converter, 199
analysis bandwidth, 208
anechoic room, 37
angle of incidence, 169–170
antinode, 37, 153
anvil, 40
articulation, 69, 122, 217
artificial head, see dummy head
associations, 219
atmospheric pressure, 9, 10
attenuation, 33–35, 137
audience, 173, 175
audiogram, 54
auditory bones, 39
authentic instruments, 68, 74, 96

B

Bach, J. S., 5, 219
band width, 34, 35
banjo, 153
barbershop singing, 100
barograph, 9
Baroque, 137, 184, 204, 215, 220, 223
base, 21
basilar membrane, 39, 41, 43, 48, 66, 68, 71
bass absorber, 174
bass reflex loudspeaker, 192
bass singer, 117, 124
bassoon, 133, 135
bass tuba, 135
bathroom, 173
bathtub, 110
beats, 30–31, 50, 51
Bebung, 152
Bel, 152
bell, 113–116, 132, 137, 165, 167–168
bellows, 140
Benade, A. H., 128, 149
Bengtsson, I., 212, 213
Bessel horn, 114
Bolin, G., 86
bone conduction, 42, 43
bow, 154–157
bow velocity, 156

bowed instruments, 143, 154–160, 204
bowing, 156
bridge, 145, 150, 154, 156, 158

C

cadence, 99, 221
cane reed instrument, 108
cantus firmus, 219
carillon bells, 167
carpets, 172, 181
carrier, 201, 202
categorical perception, 63–65
categories, perceptual, 222
CD record, 198–199
celesta, 163
cent, 46, 92–94
choir, 52, 124, 183–184
Chowning, J., 201
chromatic charge, 224
circle of fifths, 86–87, 224
clapper, 167
clarinet, 109, 133–136, 201
Classical music, 223
clavichord, 145, 152
Clynes, M., 220, 223
cochlea, 41, 66
coil, 189, 192
combination tone, 69–70
common partials, 29, 84
complex tone, 24–29, 46, 59–61, 66
compliance, 32, 33, 109
compression/rarefication, 9, 33, 40
computer, 2, 98, 209, 212, 213
concrete, 172
condenser, 189, 192
condenser microphone, 189
consonance/dissonance, 72–74, 83–86, 96, 99
cornet, 131, 135
counterphase, 11, 192
crescendo, 217
crickets, 56
critical bandwidth, 60–61, 66, 67, 70–75, 84–85, 135, 177–178
crystal pickup, 189
curtains, 172
cut-off frequency, 114, 133, 137
cycles per second, see hertz
cymbal, 165

D

Darwin, 67
dB, 20–24, 53, 98
Debussy, C., 5, 75
deafness, temporary, 54
decibel, see dB
Dekan, K., 59
denominator, 79, 97–98
difference tone distortion, 190
difference limen (DL), 49–51, 61, 102
differentiation, 222
digital recording, 198–199
digital-analog-converter, 199
directivity for microphones, 191
directional hearing, 10, 75–77, 181–182
discotheque, 53
distance, 19, 76, 171
dissonance, see consonance
distortion, 190, 191, 192, 199
DOLBY, 196
dominant, 88
Doppler effect, 16–17
drop-out, 197
drum, 164, 166
dummy head, 199–200
duodeciem, 168
Dutch tiles, 173
dyad, 84, 94, 98, 101–102

E

ear, 39–43
ear canal, 39
ear pain, 53
eardrum, 8, 38–40
echo, 170–174
edge flute, 137
electret microphone, 190
electric guitar, 154
electroacoustics, 188–189
electroglottograph, 206
electrostatic loudspeakers, 190
electro-acoustic conversion, 189–190
electro-acoustic music, 99
electrodynamic microphones, 189
electro-magnetic conversion, 189
electronic music, see electroacoustic music
embouchure, 111, 137, 138, 140
emphasis, 224–226

Index

end correction, 112
equal-loudness curves, 57
equally tempered tuning, 89–92, 94–96
erase head, 194
Eustachian tube, 39, 40, 41
expectation, 217
exponent, 90
exponential horn, 114

F

f-hole, 158
feedback, 107
feedback instruments, 128–140
filter, 201
final lengthening, 223
fission, 214, 217
flow phonation, 118
flute, 99, 109, 111, 137, 140
FM synthesis, 201, 202
FM tape recorder, 195
Fonagy, I., 226
forerunner, in piano tone outset, 150–152
formant (see also quasi-formant), 119, 121, 123, 203
Fourier analysis, 27
fraction, 80
French horn, 137
frequency, 10, 12, 15
frequency ratio, 80–105
Frydén, Lars, 222
fundamental, 25
fundamental frequency, 28, 133
fundamental frequency ratio, see frequency ratio
fusion, 214

G

Gabrielsson, I., 223
gallery, 181
gesture, 220
glass, 173
glottis, 117
gong, 165, 167
grammophone, see phonograph
grand piano, see piano
groove, 197
grouping, 96, 214, 216, 218, 222
guitar, 75, 153

H

half-wave resonator, 110, 134, 142, 143, 161
hammer, 40
hammer, in piano, 146, 149, 150
Hammond organ, 201
harmonic charge, 224
harmonic distortion, 190
harmonic interval, see just interval
harmonics, 25
harmonic series, 27
harmonic spectrum, 71, 73, 83
harmonic tempo, 179
harmonium, 128
Harmoniestärke, see harmony strength
harmony strength, 215
harp, 75, 217
harpsichord, 74, 104, 152
head, 77
hearing, 44–77
hearing loss, 55–56
hearing threshold, 53–57, 76
helicotrema, 39, 41
von Helmholtz, H., 1, 188, 192
hemiola, 221
hertz (Hz), 10
hierarchical structure, 222–223
horse power, 19
hum note, 168
Hutchins, C. M., 154

I

inharmonicity, 99, 111, 146–149, 164
input admittance, 157–158
input impedance, 128–130, 132, 139
intelligibility, 127, 179, 180, 185
intensity, 18–20, 24, 45–46, 61
intensity ratios, 24
intermodulation, 190, 201
internal spectrum, 133, 137
interval, 63, 72–74, 78–105
ionophone, 209

J

Jansson, Erick, 154
jaw opening, 122, 125, 126
JND, see difference limen
just interval, 82

just tuning, 87–89, 94–96

K

kettle drum, 164, 166
keyboard instrument, 89, 99
Krumhansl, C., 217, 224

L

ladies' choir, 124
laser light, 159
length correction, see end correction
level, 22, 53
level recorder, 205, 206
level, see sound (pressures) level
lip opening, 122
lip reed instrument, 108, 132
logarithm, 21, 22, 53, 97
longitudinal propagation, 13
losses, 33
loudness, 10, 52–53, 58–59, 61, 64
loudness level, 57–58
loudness summation, 59–61
loudness, vocal, 118
loudspeaker, 9, 77, 185, 192–194, 209
lute, 75, 153

M

macrophone, 186
magnetic field, 189, 192, 195
mallet, 166
mandolin, 153
Makeig, S., 104
marimba, 163
masked threshold, 65
masking, 65–69, 182
mass, 32, 33, 109, 141, 143, 146
Mathews, M., 99
McAdams, S., 218
mean tuning temperament of Schlick, 96
meaningful deviations, 221–222
measurement equipment, 205–210
mel, 47, 124
melodic charge, 104, 224
melodic interval, 84, 94
melograph, 206
membrane, 164–165
membrane instrument, 164–167
meridian, 167

micropause, 69, 213, 217, 224
microphones, 190–192
MIDI (Music Instrument Digital Interface), 203, 222
mineral wool, 172–173
mingograph, 205
minor tonality, 219
mirror source, 30
mistuned consonant, 30
modulation, 201
modulation transfer, 177
monophonic recording, 37
motive, 214
mouth organ, 128
mouthpiece, 114–116
muscle memory, 52
music boxes, 161, 163
music sciences, 2–4
musicology, 2
mute, 158

N

natural sciences, 4–7
natural interval, see just interval
Newton/m2 (N/m2), 10
nicks, 140
node (see also standing wave), 36, 37
nodal circles, 167
no-feedback instruments, 116–128
noise, 8, 75, 121, 122, 140, 196, 199
noise reduction system, 196
nominal note, 168
van Noorden, L., 214
note value, 63
numerator, 79, 97–98

O

oboe, 50–52, 75, 109, 133, 135
obstacles, 17–18, 169
onset/decay of tone, 34, 35, 75, 76, 133, 140, 150, 151, 217
orchestra, 55
organ, 68, 75, 104, 137, 139, 140, 181, 204, 216, 217
organ of Corti, 41, 42
oscillation, simple, 9–12
oscillations per second, see Hz

Index

oscillograph, 205
oscilloscope, 205
outer ear, see pinna
oval window, 39, 40, 41
overtone, 25, 27, 31, 133
overblowing, 131, 132
overloading, 195

P

PA, 10, 18, 117
Parsifal, 167
partials, 25
particle velocity, 15
Pascal, 10
pelog, 96
performance of music, 221
period, 8
period time, 10, 14, 15
periodic oscillation, 10
phase, 10, 25, 29, 30, 31, 76; see also counter-phase
phase angle, 11
phone, see hearing level
phonograph, 197
phrase, 222–223, 224
piano, 10, 74, 75, 104, 143, 145–152, 204
piano string (see also string and piano), 34, 143
piano touch, 149
pickup, 154
Pierce, J., 99, 216
piezo-electric crystals, 189
pinna, 39, 76
pitch, 8, 10, 44–52
pizzicato, 34
plane wave, 16
playback head, 194
podium, 181–183
pop musician, 55
power, 19, 20, 24
precedence effect, 77
pressure ratio, 24
pressure wave, 13
program music, 219
pulput, 181
pulsating sphere, 16
pure interval, see just interval
Pythagoras, 1
Pythagoran comma, 87, 94

Pythagoran tuning, 86–87, 92, 94–96, 101–103

Q

quality, perceptual, 44
quarter-wave resonator, 112, 134
quasi-formant, 134, 138

R

radiation, 16, 205
rarefaction, see compression
ratio, 78–80
Raumlinienstärke, see spatial line strength
recorder, 137
record head, 194, 195
reed, 108, 109, 128
reference, 23
reflection, 31
reflection angle, 31
reflexes, timetable for, 174, 179–181
regal, 128
relative pitch, 52
Renaissance, 5, 184
resonance, 31–33, see also formant
resonance box, 144–145
resonance box, see corpus
resonance frequencies, 32
    membranes, 164–165
    rods, 161–163
    strings, 141–146
    tubes, 108–115
resonator, see also resonance (frequency), 34, 36, 39, 108–115
reverberation, 170–171, 179
reverberation radius, 176–177
reverberation time, 171, 173–176, 185
Rimsky-Korsakow, N., 5
RMS value, 10
rod, 142, 146
rod instrument, 161–164
Romantic music, 215
Rösler, E. K., 214–215
rough timbre, 71, 84
round window, 39, 41

S

SPL (sound pressure level), 23, 57–58
Sabine's formula, 175

sampling, 198
saw tooth curve, 156, 157
saxophone, 133, 134
scala tympani, 39, 41
scala vestibuli, 39, 41
scale, 78–105
Schubert, F., 224
Seashore, C., 2, 212
self/others-balance, 183–184
semitone, 46, 63, 91, 98
serpent, 131
Shackford, C., 101
sideband, 202
simple oscillation, see oscillation and sine wave
simple tone, see sine tone
simultaneous interval, see dyad
sine wave, 12
sine tone, 9, 12
singer's formant, 124–125, 216
singing voice, 75, 116–127, 203–204
scales, 78–105
slendro, 96
slide-rule, 22
smooth timbre, 71, 72
sone, 58–59, see also loudness
sonagraph, 208, 209
soprano, 118, 125–127
sound level, 23
sound post, 156
sound pressure, 9, 10, 11, 24, 61
sound pressure ratio, 24
sound propagation, 12–16, 18
soundboard, 144–146, 153
sound sp, 15, 130
sound wave, 13, 14
spatial line strength, 215
spectrogram, 27
spectrum, 70–75
spectrum analysis, 27, 37, 42
spectrum analyzer, 208
speech, 53, 56
speed of sound, 15–16
spherical wave, 16
standing wave, 16
stapes, 40
steel pan, 165, 167
stereophonic recording, 181
streaming, 214
stretching of intervals, 103–105, 148–149

striking point, 145, 166
string, 141–144
string bowed, 142, 143
   struck, 143, 145
   plucked, 152–154
string instruments, 141–160
string quartet, 185
Stumpf, F., 188
Subbas, 16, 72
subdominant, 88
successive interval, see melodic interval
Sundberg, J., 217
synchronization, 182–183
synthesis, 201
synthesizer, 200–205
syntonic comma, 87

T

tabla, 165
tambourine, 164, 166
tape echo, 195
tape recorder, 194–197
tape speed, 196
telephone, 19
temperament, 78–105
temperature, 15, 120, 131, 145
tempo, 220
tenor, 117, 124, 126
tension, 141–144, 152
Terhardt, E., 105
threshold of hearing, 53–56
thumbtacks, 146
timbre, 70–75, 122–123
tinny old piano, 146
tone, 8
tone duration, 63, 213, 214
tone glue, 214–219
tone height, 46–48
tone wall, 186
tongue shape, 122
tongue tip, 122
tonic, 88
traffic noise, 53
transposition, 89, 99
transversal propagation, 13
traveling wave, 41
trombone, 131, 135, 137
trumpet, 115, 133, 135

Index

tube, 33, 108–115
tube ending, mynning, see also end correction, 33
tuning, 78–105, 213
tuning pipe, 30
tuning fork, 46, 50, 161, 162
tympani, see eardrum

U

una corda, 150

V

valve, oscillating, 106–107, 116, 128, 129, 131
varnish, 159
vault, 174
vestibular apparatus, 41
vibrato, 31, 51, 75, 101, 127, 201, 218
violin, 75, 154–158
violin string, 34, 154–160
viscosity losses, 111
vocal tract, 32, 119, 121, 122, 124, 125
voice source, 117, 133
voice, see singing voice
voice quality, 123
voiceprint, 208
vowel quality, 123, 127

vox humana, 128

W

Wagner, R., 167, 221
walkman, 55
wardrobe, 173
Watt, 19
wavelength, 13–18, 37, 110, 112, 142, 143, 169, 192
wave propagation, speed of, 143, 144
waveform, 25–28, 119, 156, 157, 164, 198, 205–206
Wessel, D., 215
whisper, 53, 118, 119
wind instruments, 106–140, 204
window, 174
wolf note, 158
wow, 197

X

xylophone, 163

Y

Young's modulus, 1623